Analytical Estimates of Structural Behavior

Analytical Estimates of Structural Behavior

Clive L. Dym | Harry E. Williams

CRC Press
Taylor & Francis Group
Boca Raton London New York

CRC Press is an imprint of the
Taylor & Francis Group, an **informa** business

CRC Press
Taylor & Francis Group
6000 Broken Sound Parkway NW, Suite 300
Boca Raton, FL 33487-2742

First issued in paperback 2019

© 2012 by Taylor & Francis Group, LLC
CRC Press is an imprint of Taylor & Francis Group, an Informa business

No claim to original U.S. Government works

ISBN-13: 978-1-4398-7089-1 (hbk)
ISBN-13: 978-0-367-38170-7 (pbk)

Library of Congress Cataloging-in-Publication Data

Dym, Clive L.
    Analytical estimates of structural behavior / Clive L. Dym and Harry E. Williams.
      p. cm.
    Includes bibliographical references and index.
    ISBN 978-1-4398-7091-4 (acid-free paper)
    1. Elasticity. 2. Structural dynamics. 3. Structural analysis (Engineering) I. Williams, Harry E. (Harry Edwin), 1930- II. Title.

TA418.D88 2012
624.1'76--dc23   2011035549

Visit the Taylor & Francis Web site at
http://www.taylorandfrancis.com

and the CRC Press Web site at
http://www.crcpress.com

In the spirit of "standing on the shoulders of giants," we dedicate this

book to our mentors (and the institutions at which we began to learn)

Anthony E. Armenàkas (Cooper Union)

Joseph Kempner (Brooklyn Polytechnic Institute)

Nicholas J. Hoff (Stanford University)

and

Richard M. Hermes (Santa Clara University)

Julian D. Cole (California Institute of Technology)

George W. Housner (California Institute of Technology)

Contents

Preface

We intend this book to explicitly return the notion of modeling to the analysis of structures by presenting an integrated approach to modeling and estimating structural behavior. The advent of computer-based approaches to structural analysis and design over the last 50 years has only accentuated the need for structural engineers to recognize that we are dealing with *models* of structures, rather than with the actual structures. Further, as tempting as it is to run innumerable computer simulations, closed-form *estimates* can be effectively used to guide and to check numerical results, as well as to confirm in clear terms physical insights and intuitions. What is truly remarkable is that the way of thinking about structures and their models that we propose is rooted in classic elementary elasticity: It depends less on advanced mathematical techniques and far more on thinking about the dimensions and magnitudes of the underlying physics.

A second reason for this book is our concern with traditional classroom approaches to structural analysis. Most introductory textbooks on structural analysis convey the subject as a collection of seemingly unrelated tools available to handle a set of relatively specific problem types. A major divide on the problem-type axis is the distinction drawn between structures that are statically determinate and those that are not. While this also logically conforms to a presentation in an order that reflects the respective degree of difficulty of application, it is often not seen by students as a coherent view of the discipline. Perhaps reflecting a long-standing split in professional affiliations, the classical approaches to structural analysis are often presented as a field entirely distinct from its logical underpinnings in mechanics, especially applied mechanics.

Finally, as noted before, the advent of the computer and its ubiquitous use in the classroom and in the design office has led structural engineering faculty to include elementary computer programs within a shrinking structural curriculum. Thus, students seem to spend more time and effort generating numbers, with less time and effort spent on understanding what meaning—if any—to attach to the numbers that are generated with these programs. This tendency has only strengthened as computers have become still more powerful. Still more unfortunate is that this approach emphasizes another growing dissonance in the education of engineering students: *Problems* in structural behavior and response continue to be formulated largely in mathematical terms, while *solutions* are increasingly sought with computer programs.

We have based this book on the premise that it is now even more important to understand basic structural *modeling,* with strong emphasis on understanding behavior and interpreting results in terms of the limitations

xi

of the models being applied. In fact, we would argue that the generation of numerical analyses for particular cases is, in the "real world," increasingly a task performed by technicians or entry-level engineers, rather than by seasoned professional engineers. As numerical analysis becomes both more common and significantly easier, those structural analysts and designers who know *which* calculations to perform, *how to validate and interpret* those calculations, and *what the subsequent results mean* will be the most highly regarded engineers. The knowledge needed to do these tasks can often be encapsulated and illustrated with the ability to obtain and properly use analytical, closed-form estimates—or, in other words, the ability to obtain and properly use "back of the envelope" models and formulas.

We note that it is more than the outline of topics that sets apart this book from others. That outline, to be described immediately, is not what we would expect to find in a first course in structural analysis. In fact, much of what we have included in Chapters 3–7 derives from articles we have published in the various research journals (see the references and bibliographies at the end of each chapter). The common theme of these articles and of Chapters 3–7 is the development of effective analytical estimates of the responses of certain structural models. So, we hope to stretch the mold of traditional approaches to structural analysis—and especially how our colleagues teach structural analysis—to emphasize and more explicitly describe the modeling process, and thus present a more conscious view of estimating and assessing structural response.

We finally note that while this book is not intended as a text for a *first* course in structural analysis, we certainly think it is accessible to advanced undergraduates as well as to graduate students and practitioners. It does not require deep knowledge of advanced structural mechanics models or techniques:

- We use the principle of minimum total potential energy to derive governing equations and boundary conditions, but those equations can be derived in other ways or even simply accepted.

- We introduce extensions of the Castigliano theorems and Rayleigh quotients for discrete systems, laying a foundation for applying them to continuous systems.

The mathematical skills that will be exercised are more about applying techniques of dimensional analysis, reasoning about physical dimensions, and reasoning about the relative sizes of mathematical terms and using appropriate expansions to determine limits and limiting behavior.

Organization

This book is organized as follows. In Chapter 1 we outline some important principles and techniques of mathematical modeling, including dimensional analysis, scaling, linearity, and balance and conservation laws. In Chapter 2 we review basic structural models, including structural supports and materials, as well as some general considerations of load paths, redundancy, determinacy, and stability. We also review there the concept of *idealization*, and we complete the chapter by bringing *discretization* under the modeling umbrella as well.

In Chapter 3 we use subsets of two-dimensional elasticity theory to reconsider two classic structural mechanics problems so as to explore how we develop and express physical intuition. First, we rederive the traditional fourth-order Euler–Bernoulli beam equation and boundary conditions and then use these results to estimate ranges of validity for beam models. Intuition issues emerge as we interpret both boundary conditions, the beam's physical parameters, and the nature of the loading—in particular, the transition from sets of concentrated loads to a uniform load. We illustrate how planar truss configurations behave as beams and use two-dimensional elasticity to derive another classical problem, the static response of pressure-loaded cylinders, and show how our physical intuitions can lead us astray.

In Chapter 4 we demonstrate how the behavior of arches under lateral load can be tracked as it varies from beam behavior at small values of an arch parameter (i.e., arches with very small rises) to purely compressive arch behavior when the arch parameter is large (i.e., for large arch rises). It is also shown that the behavior "flips" when the load applied is axial, rather than lateral.

In Chapter 5 we introduce two methods of analyzing coupled discrete systems, in part to lay a foundation for their application to continuous systems in our two final chapters, and in part just to ensure a common background for readers who may not be familiar with either or both of the techniques described. First, we describe recently developed extensions of Castigliano's theorems, and then we introduce Rayleigh's quotient for estimating the fundamental frequencies of coupled spring-mass oscillators. Then, in Chapter 6 we apply the extension of Castigliano's second theorem to derive simple, yet quite accurate estimates of the transverse displacements of structures modeled in terms of coupled Timoshenko beams (e.g., tall buildings). Finally, in a similar vein, in Chapter 7 we use Rayleigh quotients to analyze the dimensional behavior of and calculate numerical values of fundamental frequencies of structures modeled in terms of Euler–Bernoulli, Timoshenko, and coupled-beam systems (e.g., again, potential models of tall buildings).

Authors

Clive L. Dym is Fletcher Jones Professor of Engineering Design and director of the Center for Design Education at Harvey Mudd College. After receiving his PhD from Stanford University, Dr. Dym held appointments at the University of Buffalo; the Institute for Defense Analyses; Carnegie Mellon University; Bolt, Beranek and Newman; and the University of Massachusetts at Amherst. He has held visiting appointments at the TECHNION-Israel Institute of Technology, the Institute for Sound and Vibration Research at Southampton, Stanford, Xerox PARC, Carnegie Mellon, Northwestern, USC, and the Singapore University of Technology and Design. Dr. Dym has authored or coauthored more than a dozen books and 90 refereed journal articles, was founding editor of the journal *Artificial Intelligence for Engineering Design, Analysis, and Manufacturing,* and has served on the editorial boards of several other journals, including the ASME's *Journal of Mechanical Design.* His primary interests are in engineering design and structural mechanics.

Dr. Dym is a fellow of the Acoustical Society of America, the American Society of Mechanical Engineers, the American Society of Civil Engineers, and the American Society for Engineering Education, and is a member of the American Academy of Mechanics. Dr. Dym's awards include the Walter L. Huber Research Prize (ASCE, 1980), the Western Electric Fund Award (ASEE, 1983), the Boeing Outstanding Educator Award (first runner-up, 2001), the Fred Merryfield Design Award (ASEE, 2002), the Joel and Ruth Spira Outstanding Design Educator Award (ASME, 2004), the Archie Higdon Distinguished Educator Award (Mechanics Division, ASEE, 2006), and the Bernard M. Gordon Prize for Innovation in Engineering and Technology Education (NAE, 2012; co-winner).

Harry E. Williams is professor emeritus of engineering at Harvey Mudd College. After receiving his PhD from the California Institute of Technology, Dr. Williams joined the research staff of the Jet Propulsion Laboratory and then joined Harvey Mudd College as one of the founding faculty of its engineering program, serving there for 40 years. He has also been a Fulbright fellow at the University of Manchester, a liaison scientist for the Office of Naval Research, and a consultant to General Dynamics, Teledyne Microelectronics, the Naval Weapons Center, Aerojet-General, and the Boeing Company. He has published widely over the years, reflecting his interests in fluid mechanics, thermoelasticity, and the mechanics of solids and structures.

1

Mathematical Modeling for Structural Analysis

Summary

The dictionary defines a *model* as "a miniature representation of something; a pattern of something to be made; an example for imitation or emulation; a description or analogy used to help visualize something (e.g., an atom) that cannot be directly observed; a system of postulates, data and inferences presented as a mathematical description of an entity or state of affairs." This definition suggests that *modeling* is an activity, a *cognitive activity* in which one thinks about and makes models to describe how devices or objects of interest behave. Thus, it is important to remember that when we describe or formulate a problem in words, draw a sketch (e.g., a free-body diagram), write down or derive a formula, and crank through to get some numbers, we are modeling something. In each of these activities we are formulating and representing a model of the problem in a *modeling language*. And as we go from words to pictures to formulas to numbers, we must be sure that we are translating our problem correctly and consistently. We have to maintain our assumptions, and at the right level of detail.

Since there are many ways in which devices and behaviors can be described—words, drawings or sketches, physical models, computer programs, or mathematical formulas—it is worth refining the foregoing dictionary definition to define a *mathematical model* as a "representation in mathematical terms" of the behavior of real devices and objects. Our primary modeling language is mathematics, so we must be able to translate fluently into and from mathematics.

Scientists use mathematical models to *describe* observed behavior or results, *explain why* that behavior and those results occurred as they did, and *predict* future behaviors or results that are as yet unseen or unmeasured. *Engineers* use mathematical models to describe and analyze objects and devices in order to predict their behavior because they are interested in *designing* devices and processes and systems. Design is a consequential activity for engineers because every new airplane or building, for example, represents a model-based prediction that the plane will fly and the building stand without dire, unanticipated consequences. Further, as practicing engineers, we must always remember that we are dealing with models of a problem—*models of reality*. Thus, if our results do not match experimental data or intuitive expectations, we may well have a model that is simply wrong. So it is especially important in engineering to ask: How are such mathematical models or representations created? How are they validated? How are they used? Is their use limited and, if so, how?

To answer these and related questions, this chapter first sets out some basic principles of mathematical modeling and then goes on to describe briefly:

- abstraction and scaling
- dimensional consistency and dimensional analysis
- conservation and balance laws
- the assumption of linear behavior

Principles of Mathematical Modeling

Mathematical modeling is a principled activity that has principles behind it as well as methods that can be successfully applied. The principles are overarching or metaprinciples that are almost philosophical in nature, and they can be phrased as questions (and answers) about modeling tasks we need to perform and their purposes. That is, builders of mathematical (and other types of) models must **identify**

a. The need for the model: **Why** is this being done?
b. The data sought: **What** information is being sought?
c. The available relevant data: What is known (i.e., What is **given?**)
d. The circumstances that apply: What can be **assumed?**
e. The governing physical principles: **How** should this model be viewed?
f. The equations that will be used, the calculations that will be made, and the answers that will result: What will the model **predict?**
g. The tests to be made to validate the model and ensure its consistency with its principles and assumptions: Are the predictions **valid?**
h. The tests to be made to verify the model and ensure its usefulness in terms of the initial reason it was done: Can the predictions be **verified?**

i. Parameter values that are not adequately known, variables that should have been included, and/or assumptions that could be removed (i.e., can an iterative "model-validate-verify-improve-predict" loop be implemented? Can the model be **improved?**)

j. What will be done with the model: How will the model be **used?**

These identified tasks and questions can also be visually portrayed (see Figure 1.1).

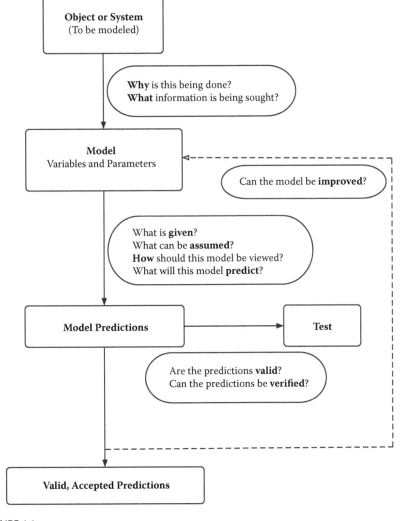

FIGURE 1.1

A graphical overview of *mathematical modeling* shows how the questions asked during a principled approach to model building relate to the development of that model. (Dym, C. L. 2004. *Principles of Mathematical Modeling,* 2nd ed. By permission of Elsevier Academic Press.)

It is worth noting that the last principle (**used?**) is often considered *early* in the modeling process, along with **why?** and **find?**, because the way a model is to be used is often intimately connected with the reason it is created. Note too that this list of questions and instructions is *not* an algorithm for building a good mathematical model. However, the underlying ideas are key to mathematical modeling, as they are key to problem formulation generally. Thus, the individual questions will recur often during the modeling process, so the list should be regarded as a general approach to *ways of thinking* about mathematical modeling.

It is most important to have a clear picture of why a model is wanted or needed. For example, a first estimate of the available power generated by a dam on a large river—say, the famed Three Gorges Dam on the Yangtze River in the People's Republic of China—would not require a model of the dam's thickness or the strength of its foundation. On the other hand, its height would be essential, as would some model and estimates of river flow quantities. By way of contrast, a design of the actual dam would need a model that incorporates all of the dam's physical characteristics (e.g., dimensions, materials, foundations) and relates them to the dam site and the river flow conditions. Thus, defining the task is the first essential step in model formulation.

The next step would be to list what is known—for example, river flow quantities and desired power levels—as a basis for listing variables or parameters that are not yet known. One should also list any relevant assumptions. For example, levels of desired power may be linked to demographic or economic data, so any assumptions made about population and economic growth should be spelled out. Assumptions about the consistency of river flows and the statistics of flooding should also be spelled out.

Which physical principles apply to this model? The mass of the river's water must be conserved, as must its momentum, as the river flows, and energy is both dissipated and redirected as water is allowed to flow through turbines in the dam (or spill over the top!). Mass must be conserved, within some undefined system boundary, because dams do accumulate water mass from flowing rivers. There are well-known equations that correspond to these physical principles. They could be used to develop an estimate of dam height as a function of power desired. The model can be validated by ensuring that all equations and calculated results have the proper dimensions, and it can be exercised against data from existing hydroelectric dams to get empirical data and validation.

If the model is inadequate or fails in some way, an *iterative loop* is then entered in which one cycles back to an earlier stage of the model building to reexamine any assumptions, known parameter values, the principles chosen, the equations used, the means of calculation, and so on. This iterative process is essential because it is the only way that models can be improved, corrected, and validated.

Abstraction and Scale (I)

Consider now issues of *scale,* of *relative size.* Size, whether absolute or relative, is very important because it affects both the form and the function of those objects or systems being modeled. Scaling influences—indeed, often controls—the way objects interact with their environments, for objects in nature, the design of experiments, or the representation of data by smooth, nice-looking curves. This section briefly discusses the ideas behind abstraction and scale, with further details to follow later.

Abstraction, Scaling, and Lumped Elements

An important decision in modeling is choosing an appropriate level of detail for the problem at hand and thus knowing what level of detail is prescribed for the attendant model. This process is called *abstraction* and it typically requires a thoughtful and organized approach to identifying those phenomena that will be emphasized—that is, to answering the fundamental question about why a model is being sought or developed. Further, thinking about finding the right level of abstraction or the right level of detail often requires finding the right *scale* for the model being developed. Stated differently, thinking about *scaling* means thinking in terms of the magnitude or size of quantities measured with respect to a standard that has the same physical dimensions.

For example, the linear elastic spring is used to model more than just the force–extension relation of simple springs such as old-fashioned butcher's scales or automobile springs. For example, $F = kx$ can be used to describe the static load-deflection behavior of a diving board, where the spring constant k will reflect the stiffness of the diving board taken as a whole, which in turn reflects more detailed properties of the board, including the material of which it is made and its own dimensions. The validity of using a linear spring to model the board can be confirmed by measuring and plotting the deflection of the board's tip as it changes with standing divers of different weights.

The classic spring equation is also used to model the static and dynamic behavior of tall buildings as they respond to wind loading and to earthquakes. These examples suggest that a simple, highly abstracted model of a building can be developed by aggregating various details within the parameters of that model. That is, the stiffness k for a building, as with the diving board, would be a *lumped element* that aggregates a great deal of information about how the building is framed, its geometry, its materials, and so on. For both the diving board and the tall building, detailed expressions of how their respective stiffnesses depended on their respective properties would be needed. It is not possible to do a detailed design of either the board or of the building without such expressions. Similarly, using springs to model atomic bonds means that their spring constants must be related to atomic interaction forces, atomic distances, subatomic particle dimensions, and so on.

Thus, the spring can be used at very large *macro* scales for buildings, at much smaller *micro* scales to model atomic bonds, and at still smaller *nano* scales to model behavior at molecular levels. The notion of scaling includes several ideas, including the effects of geometry on scale, the relationship of function to scale, and the role of size in determining limits—all of which are needed to choose the right scale for a model in relation to the "reality" we want to capture.

Another facet of the abstraction process occurs whenever, for example, a statement is made that, for some well-defined purposes, a "real," three-dimensional object behaves like a simple spring. Thus, the concept of a *lumped element* model is introduced wherein the actual physical properties of some real object or device are aggregated or *lumped* into a less detailed, more abstract expression.

An airplane, for example, can be modeled in very different ways, depending on the modeling goals. To lay out a flight plan or trajectory, the airplane can simply be considered as a point mass moving with respect to a spherical coordinate system. The mass of the point can simply be taken as the total mass of the plane, and the effect of the surrounding atmosphere can also be modeled by expressing the retarding drag force as acting on the mass point itself with a magnitude related to the relative speed at which the mass is moving. To model and analyze the more immediate, more local effects of the movement of air over the plane's wings, a model would be built to account for the wing's surface area and be complex enough to incorporate the aerodynamics that occur in different flight regimes. To model and design the flaps used to control the plane's ascent and descent, a model would be developed to include a system for controlling the flaps and also to account for the dynamics of the wing's strength and vibration response.

Clearly, a discussion about finding the right level of abstraction or the right level of detail is also a discussion about finding the right *scale* for the model being developed. *Scaling* or imposing a scale might mean assessing the effects of geometry on scale, the relationship of function to scale, and the role of size in determining limits. Each of these ideas must be addressed when the determination is made on how to scale a model in relation to the "reality" that is being captured. Since determining scale requires some sense of measurement, whether of distance, time, mass, or any property or feature, both dimensions and corresponding units must be introduced to give that measurement meaning. For example, counting an elapsed time in milliseconds, hours, or millennia can yield staggeringly different numbers, depending on the units in which that elapsed time was measured. Thus, systems of units are needed to give meaning to the measurements of dimensions.

Dimensional Consistency and Dimensional Analysis

There is a very powerful idea that is central to mathematical modeling: Every equation used must be *dimensionally homogeneous* or *dimensionally consistent;* that is, every independent, additive term in an equation must have the same net physical dimension(s). Thus, a balance of mass should have a net dimension of mass, and when forces are summed to ensure equilibrium, every independent term in that summation must have a net physical dimension of force. Dimensionally consistent equations are also called *rational* equations. In fact, ensuring that equations are (dimensionally) rational is a very useful way to validate newly developed mathematical models or to confirm formulas and equations before doing calculations with them. Beyond validation, we can go further and introduce the technique of *dimensional analysis* by which we can deduce information about a phenomenon just from the single premise that the phenomenon can be described by a dimensionally correct equation among certain variables. Some of the available tools of dimensional analysis are now described.

Dimensions and Units

The physical quantities used to model objects or systems represent *concepts,* such as time, length, and mass, to which are also attached *numerical* values or measurements. If the width of a soccer field is said to be 60 meters, the concept invoked is length or distance, and the numerical measure is 60 meters. A numerical measure implies a comparison with a standard that enables (1) communication about and (2) comparison of objects or phenomena without their being in the same place. In other words, common measures provide a frame of reference for making comparisons.

The physical quantities used to describe or model a problem are either *fundamental* or *primary* quantities, or they are *derived* quantities. A quantity is fundamental if it can be assigned a measurement standard independent of that chosen for the other fundamental quantities. In mechanical problems, for example, mass, length, and time are generally taken as the fundamental mechanical variables, while force is derived from Newton's law of motion. For any given problem, enough fundamental quantities are required to express each derived quantity in terms of these primary quantities.

The word *dimension* is used to relate a derived quantity to the fundamental quantities selected for a particular model. If mass, length, and time are chosen as primary quantities, then the dimensions of area are (length)^2, of mass density are mass/(length)^3, and of force are (mass × length)/(time)^2. The notation of brackets [ ] is introduced to read as "the dimensions of." If M, L, and T stand for mass, length, and time, respectively, then

$$[A = \text{area}] = (L)^2, \quad [\rho = \text{density}] = M/(L)^3, \quad [F = \text{force}] = (M \times L)/(T)^2 \quad (1.1)$$

The *units* of a quantity are the numerical aspects of a quantity's dimensions expressed in terms of a given physical standard. By definition, then,
a unit is an arbitrary multiple or fraction of that standard. The most widely
accepted international standard for measuring length is the meter (m), but
length can also be measured in units of centimeters (1 cm = 0.01 m) or of feet
(0.3048 m). The magnitude or size of the attached number obviously depends
on the unit chosen, and this dependence often suggests a choice of units to
facilitate calculation or communication. For example, a soccer field width can
be said to be 60 m, 6000 cm, or (approximately) 197 ft.

Dimensions and units are related by the fact that identifying a quantity's
dimensions allows us to compute its numerical measures in different sets
of units, as we just did for the soccer field width. Since the physical dimensions of a quantity are the same, there must exist numerical relationships
between the different systems of units used to measure the amounts of that
quantity—for example, 1 foot (ft) $\cong$ 30.48 centimeters (cm), and 1 hour (hr) =
60 minutes (min) = 3,600 seconds (sec *or* s). This equality of units for a given
dimension allows units to be changed or converted with a straightforward
calculation—for example,

$$65\frac{\text{mi}}{\text{hr}} = 65\frac{\text{mi}}{\text{hr}} \times 5280\frac{\text{ft}}{\text{mi}} \times .3048\frac{\text{m}}{\text{ft}} \times .001\frac{\text{km}}{\text{m}} \cong 104.6\frac{\text{km}}{\text{hr}} \qquad (1.2)$$

Each of the multipliers in this conversion equation has an effective value
of unity because of the equivalencies of the various units; that is, 1 mi =
5280 ft, and so on. This, in turn, follows from the fact that the numerator and
denominator of each of the preceding multipliers have the same physical
dimensions.

Dimensionally Homogeneous Equations and Unit-Specific Formulas

By definition, a *rational equation* is dimensionally homogeneous, which
means each independent, additive term in that equation has the same net
dimensions. Simply put, length cannot be added to area, or mass to time, or
charge to stiffness. Quantities having the same dimensions but expressed in
different units can be added, although with great care (e.g., length in meters
and length in feet). The fact that equations must be rational in terms of their
dimensions is central to modeling because it is one of the best—and easi-
est—checks to make to determine whether a model makes sense, has been
correctly derived, or even has been correctly copied!

A *dimensionally homogeneous equation* is independent of the units of measurement used. However, unit-dependent versions of such equations are
often created for doing repeated calculations or as a memory aid. Consider
the well-known hydrostatic condition as applied to a traditional liquid
manometer. For constant specific weight of the manometer liquid, γ, the local

atmospheric pressure can be written in terms of the height, h, of the liquid column in the manometer:

$$p_a = \gamma h \qquad (1.3)$$

Equation (1.3) is dimensionally homogeneous because the net physical dimensions of γh are F/L^2 (force/length^2), which are the same as those of the pressure, p_a. Thus, Equation (1.3) is independent of the system of units in which the height, specific weight, and pressure are measured. Weather forecasters used to measure the local atmospheric pressure with mercury manometers. Since $\gamma_M = 133{,}100$ N/m^3, the height of the mercury column that denotes an atmospheric pressure p_a is simply

$$h = \frac{p_a}{\gamma_M} \equiv \frac{p_a}{133{,}100} \text{ m} \qquad (1.4)$$

Equation (1.4) is not dimensionally homogeneous because it assumes a particular value—both numbers *and* units—for the specific weight. Then the atmospheric pressure is measured in those same units. Similarly, were the manometer to be a column of seawater, since $\gamma_W = 65$ lb/ft^3, the corresponding manometer height would be

$$h = \frac{p_a}{\gamma_W} \equiv \frac{p_a}{65} \text{ ft} \qquad (1.5)$$

An interesting aside to this discussion of dimensional homogeneity is that the practical reporting of pressure levels is a mess! The atmospheric pressure at sea level is $p_a = 101{,}350$ N/m^2, for which Equation (1.4) yields a value of $h = 0.761$ m. A weatherperson would report this as 29.96 in. Hg (inches of mercury) because the manometer likely has traditional American units scribed on its tube. It is also of note that this simple analysis of dimensions and units explains why mercury manometers are used: Equation (1.5) shows that a seawater manometer would have to be 32.6 ft high to measure a pressure of 14.7 psi. To add to the confusion, scuba divers refer to pressure in *atmospheres*: one atmosphere equals the standard sea-level value of the atmospheric pressure, $p_a \triangleq 101{,}350$ N/m$^2 \equiv$ 1 atm. Then, at a depth $h = 32.4$ ft below the sea surface, a diver would report a pressure reading of 2 atm. (Why not 1 atm?) Remember, though, that Equations (1.4) and (1.5) are *unit-specific formulas*. So, while these formulas may be useful or even elegant, their ranges of validity are strictly limited.

The Basic Method of Dimensional Analysis

Dimensional analysis is the technique used to ensure dimensional consistency. First, the dimensions of all derived quantities are checked to see that they are properly represented in terms of the chosen primary quantities and their dimensions. Second, the proper *dimensionless groups* of variables—ratios and products of problem variables and parameters that are themselves dimensionless—are identified. There are two different techniques for identifying such dimensionless groups: the *basic method* and the *Buckingham pi theorem*.

The basic method of dimensional analysis is a rather informal, unstructured approach for determining dimensional groups. It depends on being able to construct a functional equation that contains all of the relevant variables, for which we know the dimensions, and then formulating dimensionless groups by thoughtfully eliminating dimensions. To illustrate the basic method, consider determining the tip deflection of a cantilever beam that has a bending stiffness $B = EI$ and length L when the beam is subjected to a tip load P and a load q_0 uniformly distributed along its length. The goal is to find a dimensionless function that relates the tip deflection, δ, to both loads and to the beam's stiffness and length—that is,

$$\delta = f_1(B, L, P, q_0) \tag{1.6}$$

The dimensions for the five variables in Equation (1.6) are

$$[\delta] = L, \quad [B] = FL^2, \quad [L] = L, \quad [P] = F, \quad [q_0] = F/L \tag{1.7}$$

Note that there are only two fundamental dimensions, F and L, in Equation (1.7), and only two of the five variables, δ and L, have the same dimensions. Thus, the basic method can be expected to identify three (i.e., $5 - 2$) groups of the five variables. The force dimension is eliminated first by dividing force-related variables; for example, starting with P,

$$\delta = f_2\left(\frac{B}{P}, \frac{q_0}{P}, L\right) \tag{1.8}$$

Then, given the net dimensions of B/P, q_0/P, and δ, it makes sense to formulate three dimensional groupings by eliminating the length as both variable and dimension:

$$\frac{\delta}{L} = f_3\left(\frac{B}{PL^2}, \frac{q_0 L}{P}\right) \tag{1.9}$$

Equation (1.9) might be considered a formal final result. However, two further considerations are warranted. Since the second dimensionless ratio in Equation (1.9) reflects a ratio of the two different loads, it is fairly easy to see that an equivalent formulation of that result is

$$\frac{\delta}{L} = f_4\left(\frac{PL^2}{B}, \frac{q_0 L^3}{B}\right) \tag{1.10}$$

Further, if the discussion is limited to the (classical) linear theory of beams, then Equation (1.10) assumes a final form that is entirely consistent with elementary beam theory—that is,

$$\frac{\delta}{L} = C_1 \left(\frac{PL^2}{B} \right) + C_2 \left(\frac{q_0 L^3}{B} \right) \tag{1.11}$$

The basic method is not unique; that is, the steps just performed could have been done differently. The elimination of the force dimension could have also been done by using B to eliminate the force dimension, so that

$$\delta = f_2' \left(\frac{P}{B}, \frac{q_0}{B}, L \right) \tag{1.12}$$

Then, eliminate the length, so that

$$\delta = f_3' \left(\frac{PL^2}{B}, \frac{q_0 L^3}{B} \right) \tag{1.13}$$

Equations (1.10) and (1.13) clearly represent the same result. The point is that the steps taken could have been implemented differently, but the end result is unchanged. In a similar vein, it is not hard to see that the force elimination leading to Equation (1.12) could have been written as

$$\delta = f_2'' \left(\frac{B}{P}, \frac{B}{q_0}, L \right) \tag{1.14}$$

and the end result would still be the same. The steps actually taken are often guided by experience and prior knowledge—for example, it is to be expected from beam theory that $\delta \sim PL^3/B$—but if care is taken in eliminating dimensions, the basic method will produce the correct results even if there is not sufficient advance knowledge to guide the process.

These applications of the basic method of dimensional analysis show that it does not have a formal algorithmic structure, but that it can be described as a series of steps to take:

1. List all of the variables of the problem and their dimensions.
2. Identify one variable as depending on the remaining variables and parameters.
3. Express that dependence in a functional equation (i.e., the analog of Equation 1.6).

4. Eliminate one of the primary dimensions to obtain a revised functional equation.

5. Repeat step 3 until a revised, *dimensionless* functional equation is found.

6. Review the final *dimensionless* functional equation to see whether the apparent behavior accords with the behavior anticipated in step 6.

The Buckingham Pi Theorem of Dimensional Analysis

Buckingham's pi theorem represents a more formal approach to dimensional analysis that can be stated as follows: A dimensionally homogeneous (or rational) equation involving n variables in m primary or fundamental dimensions can be reduced to a single relationship among $n - m$ independent dimensionless products. This means that any one term in the equation can be defined as a function of all of the others. If Buckingham's Π notation is introduced to represent a dimensionless term, his famous pi theorem can be written as

$$\Pi_1 = \Phi(\Pi_2, \Pi_3 \ldots \Pi_{n-m}) \tag{1.15a}$$

or, equivalently,

$$\Phi(\Pi_1, \Pi_2, \Pi_3 \ldots \Pi_{n-m}) = 0 \tag{1.15b}$$

Equations (1.15a) and (1.15b) state that a problem with n derived variables and m primary dimensions or variables requires $n - m$ dimensionless groups to correlate all of its variables.

The pi theorem is applied by first identifying the n derived variables in a problem: $A_1, A_2, \ldots A_n$. Then m of these derived variables are chosen such that they contain all of the m primary dimensions, say, A_1, A_2, A_3 for $m = 3$. Dimensionless groups are then formed by permuting each of the remaining $n - m$ variables $(A_4, A_5, \ldots A_n$ for $m = 3)$ in turn with those m variables already chosen:

$$\Pi_1 = A_1^{a_1} A_2^{b_1} A_3^{c_1} A_4$$
$$\Pi_2 = A_1^{a_2} A_2^{b_2} A_3^{c_2} A_5$$
$$\vdots \tag{1.16}$$
$$\Pi_{n-m} = A_1^{a_{n-m}} A_2^{b_{n-m}} A_3^{c_{n-m}} A_n$$

The a_i, b_i, and c_i are chosen to make each of the permuted groups Π_i dimensionless.

Consider once again the loaded cantilever to illustrate how Buckingham's pi theorem is applied. The five variables and their two fundamental dimensions were given in Equation (1.7). In this case $m = 5$ and $n = 2$, so once again three dimensionless groups are expected. If L and B are chosen as the variables around which to permute the remaining two variables (δ, P, q_0) to obtain the three groups, it follows that

$$\Pi_1 = L^{a_1} B^{b_1} \delta$$

$$\Pi_2 = L^{a_2} B^{b_2} P \qquad (1.17)$$

$$\Pi_3 = L^{a_3} B^{b_3} q_0$$

The pi theorem applied here then yields three dimensionless groups:

$$\Pi_1 = \frac{\delta}{L}$$

$$\Pi_2 = \frac{PL^2}{B} \qquad (1.18)$$

$$\Pi_3 = \frac{q_0 L^3}{B}$$

These groups are exactly the same as those found before, using the basic method.

One of the "rules" of applying the pi theorem is that the m chosen variables include all n of the fundamental dimensions, but no other restrictions are given. So, it is natural to ask how this analysis would change if one started with three different variables. For example, suppose L and P were chosen as the variables around which to permute the remaining three variables (δ, B, q_0) to obtain the three groups. In this case,

$$\Pi_1' = L^{a_1} P^{b_1} \delta$$

$$\Pi_2' = L^{a_2} P^{b_2} B \qquad (1.19)$$

$$\Pi_3' = L^{a_3} P^{b_3} q_0$$

This application of the pi theorem to Equation (1.19) yields the following three "new" dimensionless groups:

$$\Pi_1' = \frac{\delta}{L}$$

$$\Pi_2' = \frac{B}{PL^2} \qquad (1.20)$$

$$\Pi_3' = \frac{q_0 L}{P}$$

Equation (1.20) presents exactly the same information as Equation (1.18), albeit in slightly different forms: $\Pi_1' \equiv \Pi_1$, $\Pi_2' \equiv 1/\Pi_2$, and $\Pi_3' \equiv \Pi_3/\Pi_2$. Thus, their inclusion of the same dimensionless groups suggests that the number of dimensionless groups is unique, but that the precise forms that these groups may take are not. These examples demonstrate that the dimensionless groups determined in any one calculation may be unique, but they may also take on different, albeit related, forms when calculated somewhat differently.

Finally, note that both the basic method and the Buckingham pi theorem can be cast in similar algorithmic structures. However, experience and insight are key to applying both methods, even for elementary problems.

Abstraction and Scale (II)

We now return to issues of scale, albeit in more detail. In particular, we now briefly discuss size and shape, size and function, scaling and conditions that are imposed at an object's boundaries, and some of the consequences of choosing scales in both theory and experimental measurements. As noted earlier, finding the right level of abstraction or the right level of detail means finding the right scale for the model being developed, which in turn means assessing the effects of geometry on scale, the relationship of function to scale, and the role of size in determining limits. These ideas must be addressed when a model is scaled in relation to the "reality" that is being captured.

The scale of things is often examined with respect to a magnitude that is set within a standard. Thus, when talking about freezing phenomena, temperatures are typically referenced to the freezing point of materials included in the model. Similarly, the models of Newtonian mechanics work extraordinarily well for virtually all earth- and space-bound applications. Why is that so? Simply because the speeds involved in all of these calculations are far, far smaller than c, the speed of light in a vacuum. Thus, even a rocket fired at escape speeds of 45,000 km/hr seems to stand still when its speed is compared to $c \approx 300{,}000$ km/s $= 1.080 \times 10^9$ km/hr!

These scaling ideas not only extend the ideas discussed earlier about dimensionless variables but also introduce the notion of *limits*. For example, in Einstein's general theory of relativity, the mass of a particle moving at speed, v, is given as a (dimensionless) fraction of the rest mass, m_0, by

$$\frac{m}{m_0} = \frac{1}{\sqrt{1 - \left(\dfrac{v}{c}\right)^2}} \tag{1.21}$$

The scaling issue here is to find the limit that supports the customary practice of taking the masses or weights of objects to be constants in everyday life and in normal engineering applications of mechanics. A box of candy is not expected to weigh any more whether one is standing still, riding in a car at 120 km/hr (75 mph), or flying across the country at 965 km/hr (600 mph). This means that the square of the dimensionless speed ratio in Equation (1.21) is much less than 1, so that $m \cong m_0$. According to Equation (1.21), for that box of candy flying across the country at 965 km/hr = 268 m/s, that factor in the denominator of the relativistic mass formula is

$$\sqrt{1-\left(\frac{v}{c}\right)^2} = \frac{m_0}{m} = \sqrt{1-7.98 \times 10^{-13}} \cong 1 - 3.99 \times 10^{-13} \cong 1 \qquad (1.22)$$

Clearly, for practical day-to-day existence, such relativistic effects can be neglected. However, it remains the case that Newtonian mechanics is a good model only on a scale where all speeds are very much smaller than the speed of light. If the ratio v/c becomes sufficiently large, the mass can no longer be taken as the constant rest mass, m_0, and Newtonian mechanics must be replaced by relativistic mechanics.

In a similar vein, modern electronic components and computers provide further evidence of how limits in different domains have changed the appearance, performance, and utility of a wide variety of devices. The bulky radios of the 1940s and 1950s, or the earliest television sets, were very large because their electronics were implemented with old-fashioned vacuum tubes. These tubes were large and threw off an enormous amount of heat energy. The wiring in these circuits looked like standard electrical wiring in a house or office building. Now, of course, people carry television sets, personal digital assistants (PDAs), and wireless telephones in their pockets. These new technologies have emerged because the limits on fabricated electrical circuits and devices have dramatically changed, as they have also on the design and manufacturing of small mechanical objects.

This is true beyond electronics. The scale at which surgery is done on people has changed because of new abilities to "see" inside the human body with greater resolution—with increasingly sophisticated scans and imagers, as well as with fiber-optic television cameras—and to design visual, electronic, and mechanical devices that can operate inside a human eye and in arteries and veins. Devices are being engineered at the molecular level in the emerging field of nanotechnology. Thus, the mathematical models will change, as will the resulting devices and "machines."

Geometric Scaling

Consider two cubes, one of which has sides of unit length in any system of units; that is, the cube's volume could be 1 in.^{3} or 1 m^3, or 1 km^3. The other cube has sides of length L in the same system of units, so its volume is either L^3 in.^{3} or L^3 m^3, or L^3 km^3. Thus, for comparison's sake, the units in which the two cubes' sides are actually measured can be ignored. The total area and volume of the first cube are, respectively, 6 and 1, while the corresponding values for the second cube are $6L^2$ and L^3. An instance of *geometric scaling* can be immediately seen; that is, the area of the second cube changes as does L^2 and its volume scales as L^3. Thus, doubling the side of a cube increases its surface area by a factor of four and its volume by a factor of eight.

Now consider the data that emerged from a study of medieval churches and cathedrals in England. Large churches and cathedrals of that area (Figure 1.2a, b) were typically laid out in a cruciform pattern (Figure 1.2c) where the *nave* was the major longitudinal area that extended from the front entrance to the *chancel* or altar area at the back. A *transept* section was set out perpendicular to the nave, close to the chancel. Was the cruciform shape inspired by religious feeling, which would not seem surprising? Or was the cruciform shape inspired by the scaling that responded to the need for both good lighting and sound structures?

(a)

FIGURE 1.2

(a) Cruciform plan of the Norwich Cathedral (after Gould, S. J. 1975. *Harvard Magazine* 78 (2): 43–50); (b) gothic nave of Canterbury Cathedral (by permission of the late S. J. Gould and Elsevier Academic Press); (c) cross section of Westminster Abbey (after Heyman, J. 1995. *The Stone Skeleton.* Cambridge, England: Cambridge University Press).

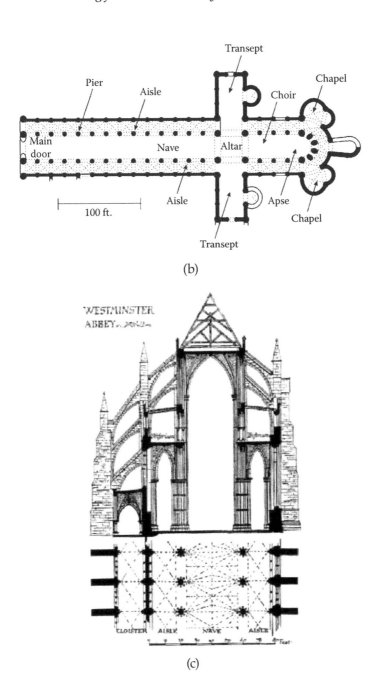

(b)

(c)

FIGURE 1.2
(Continued)

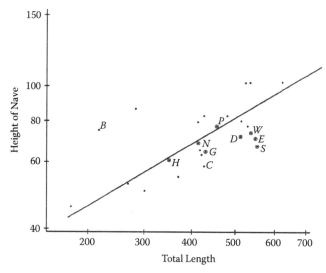

FIGURE 1.3
Plot of log (nave height) against log (church length) with dots representing various Romanesque churches and capitalized letters representing English churches as follows: B, Earl's Barton; C, Chichester; D, Durham; E, Ely; G, Gloucester; H, Hereford; N, Norwich; P, Peterborough; S, St. Albans; W, Winchester (by permission of the late S. J. Gould and Elsevier Academic Press).

We start answering these questions by taking the length of a church as the first-order indicator of its size. Thus, the longer its length is, the larger the church is. Figure 1.3 displays a *log–log* plot of nave height against church length for a variety of medieval cathedrals and churches in England and on the European continent. We note that as church lengths (and thus size) increase, the nave heights increase in absolute terms but fall off in *relative* terms. That is, as churches get longer (and bigger), their naves get relatively smaller. Further, although we do not give the data to buttress this assertion, the larger churches tend to have *narrower* naves. Why do the nave height and width not increase with church size? The answer lies in the geometric scaling of surface areas and enclosed volumes.

The relevant scaling issue is the change of a church's enclosed area as it is made longer (and larger). A longer church has a longer perimeter. In buildings of constant shape, the surface area of the enclosing wall increases linearly with the perimeter length, while the enclosed volume increases with the square of the perimeter length. And it becomes more difficult for light and fresh air to penetrate into the church's interior as its perimeter increases. (Remember that these marvelous structures were built long before the invention of the lightbulb and air conditioning!) However, the severity of the lighting and ventilation problems can be mitigated by introducing a transept because it enables a relatively constant nave width, thus taking away the "constant shape" constraint. If the width is kept constant, then the enclosed area increases only *linearly* with perimeter length, along with the church's length and size. Then, of course, such a church will be seen as relatively narrow!

The enclosed area can be increased by widening the nave, but this also exacerbates interior lighting and ventilation problems. And it creates still another problem—namely, that of building a larger roof to cover the larger enclosed area. Since roofs of cathedrals and churches were built to sit atop stone *vaults* and *arches*, roof spans became a critical design issue because it was very hard to build wide stone arches and vaults. Further, building wide arches also restricts the height of the nave because it is the nave walls that support the outward thrust developed in the roof vaults, even when the nave walls are supported by flying buttresses (Figure 1.2c). Thus, higher nave walls had to be made thicker to support both their own weight and the weight of the roofs supported by the vaults, which were in turn supported at the more flexible tops of the walls. That is why the width and height of cathedral naves had to be scaled back as overall church length (size) was increased to solve problems of lighting, ventilation, *and* structural safety.

As a footnote to this discussion, note that the log–log plot of Figure 1.3 is a very handy way to graph large ranges of data. Of course, modern computational capabilities enable us to skip the "old fashioned" and laborious plotting of data by simply entering tables of data points so that a computer generates equations or curves. But something is gained by thinking through these issues without a computer.

Scale in Equations: Size and Limits

We have noted before—and seen, in the relativistic mass example—that limits occur quite often in mathematical modeling. They may control the size and shape of an object, the number, kinds of variables, and the range of validity of an equation, or even the application of particular physical models—or "laws," as they are often called.

In certain situations, scaling may shift limits or points on an object's boundary where *boundary conditions* are applied. For example, consider the hyperbolic sine,

$$\sinh x = \frac{1}{2}(e^x - e^{-x}) \tag{1.23}$$

For large values of x, the term e^x will be much larger than the term e^{-x}. The approximation problem is one of defining an appropriate criterion for discarding the smaller term, e^{-x}. For dimensionless values of x greater than 3, the second term on the right-hand side of Equation (1.23), e^{-x}, does become very small (less than 4.98×10^{-2}) compared to e^x for $x = 3$, which is 20.09. Hence, one could generally take $\sinh x \cong \frac{1}{2} e^x$. All that must be decided is a value of x for which the approximation $e^{2x} - 1 \cong e^{2x}$ is acceptable.

This problem can be approached by introducing a *scale factor,* λ, which can be used to look for values of x for which the approximation

$$\sinh(x/\lambda) \cong \frac{1}{2}e^{x/\lambda} \qquad (1.24)$$

can be made. Putting a scale factor, λ, in the approximation of Equation (1.24) obviously means that it will affect the value of x for which that approximation is acceptable. Now the comparison is one that wants

$$e^{2x/\lambda} - 1 \cong e^{2x/\lambda}. \qquad (1.25)$$

For $\lambda = 1$, the approximation is good for $x \geq 3$, while for $\lambda = 5$, the approximation works for $x \geq 15$. Thus, by introducing the scale factor λ, we can make the approximation valid for different values of x because we are now saying that $e^{-x/\lambda}$ is sufficiently small for $x/\lambda \geq 3$. Changing λ has in effect changed a boundary condition because it has changed the expression of the boundary beyond which the approximation is acceptable to $x \geq 3\lambda$.

Recall that functions such as the exponentials of Equations (1.24) and (1.25), as well as sinuosoids and logarithms, are *transcendental functions* that can always be represented as power series. For example, the Taylor series for the exponential function is

$$e^{x/\lambda} = 1 + \frac{x}{\lambda} + \frac{1}{2!}\left(\frac{x}{\lambda}\right)^2 + \frac{1}{3!}\left(\frac{x}{\lambda}\right)^3 + \cdots + \frac{1}{n!}\left(\frac{x}{\lambda}\right)^n + \cdots \qquad (1.26)$$

It is clear from here that the argument of the exponential must be dimensionless because, without this property, Equation (1.26) would not be a rational equation. Further, one could not calculate numerical values for the exponential—or any other transcendental—function, if its argument was not dimensionless. The scale factor in Equation (1.26) makes the exponential's argument dimensionless and thus enables numerical calculations.

We will model only Hookean materials whose stress–strain laws are linear (i.e., $\sigma = E\varepsilon$). However, many materials exhibit *viscoelastic* behavior in which their response combines Hookean elasticity with Newtonian viscosity, for which the stress is linearly proportional to the strain rate (i.e., $\sigma = \mu\dot{\varepsilon}$). When a Hookean element is combined with a Newtonian element in series so that the stress in each is the same while their strains are added (see Figure 1.4), the equation of equilibrium is

$$\frac{d\sigma(t)}{dt} + \frac{E}{\mu}\sigma(t) = E\frac{d\varepsilon(t)}{dt} \qquad (1.27)$$

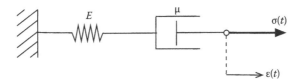

FIGURE 1.4
A Maxwell model of a viscoelastic system in which linear Hookean (i.e., $\sigma(t) = E\varepsilon(t)$) and Newtonian (i.e., $\sigma(t) = \mu\dot{\varepsilon}(t)$) elements are connected in series.

If such a system were held at a constant strain level, $\varepsilon(t) = \varepsilon_0 H(t)$, where $H(t)$ is the unit step function, then Equation (1.27) becomes

$$\frac{d\sigma(t)}{\sigma(t)} = -\frac{E}{\mu}dt. \tag{1.28}$$

Equation (1.28) is termed a *rate equation* because it shows the rate at which the stress changes its value as time changes. Further, that rate equation will be a rational equation only if the net dimensions of each side of Equation (1.28) are the same, which means that each side must be dimensionless. The left-hand side is clearly dimensionless because it is the ratio of a stress change to the stress itself. The right-hand side is dimensionless *only* if the factor μ/E has physical dimension of time, T, which is the case because for the elastic constant $[E] = FL^{-2}$ and for the coefficient of viscosity $[\mu] = FTL^{-2}$. Then the ratio $\tau \triangleq \mu/E$ can be seen as the *characteristic time* for this Maxwell model, which can also be seen as a unit in which to measure the time it takes for the stress to decrease to some specified fraction of its initial value.

The formal solution to Equation (1.28) can then be expressed in terms of τ as

$$\sigma(t) = \sigma_0 e^{-t/\tau} \tag{1.29}$$

where $\sigma_0 = \sigma(0)$ is the initial value of the stress dictated by the imposition of the strain step load. If a *relaxation time* $t_{relaxation}$ is then formally defined as that amount of time required for the stress to fall to $1/e$ of its initial value— that is,

$$\sigma(t_{relaxation}) \triangleq \frac{\sigma_0}{e} = \sigma_0 e^{-t_{relaxation}/\tau} \tag{1.30}$$

that relaxation time follows from Equation (1.30) as

$$t_{relaxation} \equiv \tau \tag{1.31}$$

Thus, by the definition (Equation 1.30), the relaxation time is equal to the viscoelastic system's characteristic time.

Consequences of Choosing a Scale

Since all actions have consequences, it should come as no surprise that the acquisition of experimental data, their interpretation, and their perceived meaning(s) generally can be very much affected by the choice of scales for presenting and organizing data. To illustrate how scaling affects data acquisition, consider the diagnosis of a malfunctioning electronic device such as an audio amplifier. Such amplifiers are designed to reproduce their electrical input signals without any distortion. The outputs are distorted when the input signal has frequency components beyond the amplifier's range or when the amplifier's power resources are exceeded. Distortion also occurs when an amplifier component fails, in which case the failure must be diagnosed to identify the particular failed component(s).

Such diagnoses are commonly obtained from oscilloscope displays of the device's output to a known input signal. If the device is working properly, a clear, smooth replication of the input would be expected. One standard test input is the square wave shown in Figure 1.5(a). One replication of that square wave is shown in Figure 1.5(b), and it seems just fine until it is noticed that the horizontal time scale is set at a fairly high value—that is, 0.5 second/division. To ensure that something that might not show up on this scale is not overlooked, we spread out the same signal on shorter time scales of 0.5 millisecond/division Figure 1.5(c) and 0.5 microsecond/division Figure 1.5(d), neither of which is a nice sinusoid. This suggests that the device is malfunctioning. Had the oscilloscope not been set to shorter, more appropriate time scales, an erroneous conclusion might have been reached.

Scaling and the Design of Experiments

Scale also affects the ways in which experiments are designed, especially when the context is that of ensuring that models replicate the prototypes or "real" artifacts that they are intended to stand for or model. This aspect of scaling is, as we will now show, intricately intertwined with our earlier discussion of dimensional analysis.

Scale models or reproductions of physical phenomena or devices are used to do experiments and study behavior to confirm or to suggest an analytical model. Often, laboratory experiments are more easily done than full-scale tests. For example, it is easier to study the vibration characteristics of a model of a proposed bridge design than it is to build the designed bridge and hope for the best, just as it is easier to test models of rockets in simulated spaceflight or models of buildings in simulated earthquakes or fires. But such experimental models are not very useful without preliminary analysis or the development of clear physical hypotheses. We illustrate that with a simple example.

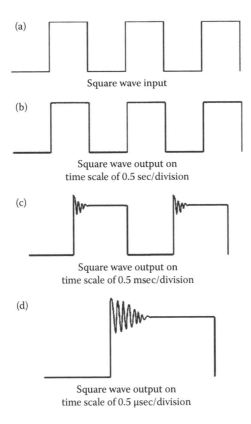

(a)

Square wave input

(b)

Square wave output on
time scale of 0.5 sec/division

(c)

Square wave output on
time scale of 0.5 msec/division

(d)

Square wave output on
time scale of 0.5 μsec/division

FIGURE 1.5
A square wave (a) is the input signal to a (hypothetical) malfunctioning electronic device. Traces of the output signals are shown at three different time scales (i.e., long, short, shorter): (b) 0.5 second/division; (c) 0.5 millisecond/division; and (d) 0.5 microsecond/division. (Dym, C. L. 2004. *Principles of Mathematical Modeling,* 2nd ed. By permission of Elsevier Academic Press.)

Consider a very long, simple beam, loaded only by its own weight per unit length w. We can see from Equation (1.20) that the three dimensionless groups for this problem are such that the tip deflection δ can be written as

$$\frac{\delta}{L} = f\left(\frac{PL^2}{B}, \frac{wL^3}{B}\right) \tag{1.32}$$

where w is now a parameter that replaces the uniform load q_0, rather than an independently chosen load.

Suppose we want to conduct an experiment with a small, light *model* beam in order to predict the behavior of a large, heavy *prototype* beam. How would we organize such an experiment? The answer begins with dimensional analysis. The model properties and prototype properties must be such that the three dimensional groups have the same numerical values for both model and prototype. Stated in mathematical terms, with subscripts m for model and p for prototype,

$$(\Pi_i)_m = (\Pi_i)_p \quad i = 1, 2, 3 \tag{1.33}$$

That is, while we can scale the geometry, material weight and stiffness, or load for our own convenience to some extent, we cannot scale *all* of the independent variables independently. In order to preserve the property of *complete similarity* between model and prototype, we must preserve the equality of each dimensionless group between model and prototype. That is, if we apply the similarity rules of Equation (1.33) to the specific case of the beam whose dimensionless groups are given in Equation (1.20), we must preserve complete similarity by requiring that

$$\left(\frac{\delta}{L}\right)_m = \left(\frac{\delta}{L}\right)_p, \quad \left(\frac{PL^2}{B}\right)_m = \left(\frac{PL^2}{B}\right)_p, \quad \left(\frac{wL^3}{B}\right)_m = \left(\frac{wL^3}{B}\right)_p \tag{1.34}$$

Having now established the overall conditions needed for complete similarity, we can now go into detail to see both what we *must* do and what we *may* do in terms of *scaling factors* defined for each of the problem's variables— that is, for the factors

$$n_\delta = \frac{\delta_p}{\delta_m}, \quad n_P = \frac{P_p}{P_m}, \quad n_w = \frac{w_p}{w_m}, \quad n_B = \frac{B_p}{B_m}, \quad n_L = \frac{L_p}{L_m} \tag{1.35}$$

Thus, we see that the scaling factors in Equation (1.35) are simply ratios of the values of each of the variables in the prototypes to the values of the same variable in the model. Equation (1.35) shows that we have five such scaling factors for this problem, while Equation (1.34) shows three overall similarity conditions that must be satisfied. However, it is crucial that we remember that one of our parameters actually aggregates two other parameters: $B = EI$. Thus, this problem appears to have six scaling factors:

$$n_\delta = \frac{\delta_p}{\delta_m}, \quad n_P = \frac{P_p}{P_m}, \quad n_w = \frac{w_p}{w_m}, \quad n_E = \frac{E_p}{E_m}, \quad n_I = \frac{I_p}{I_m}, \quad n_L = \frac{L_p}{L_m} \tag{1.36}$$

We can then write the similarity conditions (Equation 1.34) in terms of this set of six scaling factors as

$$n_\delta = n_L, \quad n_P n_L^2 = n_E n_I, \quad n_w n_L^3 = n_E n_I \tag{1.37}$$

So, if we choose a length scale (n_L) for this problem, we have also chosen a deflection scale (n_δ) by the first part of Equation (1.37). However, this means that we may still freely choose four of the remaining scaling factors (n_P, n_w, n_E, and n_I). If we choose the scaling factors of the elastic modulus (n_E) and of the weight (n_w) because we have appropriate materials and/or small beams lying around our laboratory, then the two remaining scaling factors, n_P and n_I, are determined by the second and third parts of Equation (1.37):

$$n_P = \frac{n_E n_I}{n_L^2}, \quad n_I = \frac{n_w n_L^3}{n_E} \tag{1.38a}$$

which in turn dictates that

$$n_P = n_w n_L \tag{1.38b}$$

Equations (1.38a) and (1.38b) suggest that the order in which model properties are chosen for experiments is not arbitrary. For example, suppose we wanted to predict the deflection of a steel prototype by doing experiments on a small model made of pine. Since the model's material has been chosen, a laboratory scenario would start with noting that

- The moduli of elasticity scale roughly as $n_E = 30 \times 10^6 \, \text{psi}/1 \times 10^6 \, \text{psi} = 30$
- The weights scale roughly as $n_w = 0.283 \, \text{lb/in.}^3/24 \, \text{lb/ft}^3 \sim 20$

The length scale is then defined by our choice of the length of the model; that is,

- Choose the model so that its length is 1/20 of the length of the prototype, so $n_L = 20$.

Then it follows that the model must be shaped or sculpted so that the second moment of its cross-sectional area satisfies Equation (1.38a), which means that

- The scaling factor of the second moment of the area is $n_i = n_w n_L^3/n_E$ ~ 5,333.

Then, for a truly similar experiment, when we apply a laboratory load to the model of $P_m = P_p/400$, we would expect to measure a laboratory deflection of $\delta_m = \delta_p/20$.

Note that this introduction to the consequences of scaling in modeling is short and rather limited. Clearly, not all experiments are so easily analyzed or scaled, and there are many more issues to be explored in scaling in the design of experiments.

Notes on Approximating: Dimensions and Numbers

This book, like many engineering textbooks, focuses on solving problems within a restricted domain—here, structural mechanics. The basic principles involved in this book are few—one might even argue, to extend an old Chassidic joke, that Newtonian mechanics is contained entirely within $\vec{F} = m\vec{a}$ and everything else is just a matter of detail—but the approach to setting up and solving structures problems taken herein is the "real stuff." There are some key modeling ideas to be emphasized that have to do with magnitudes and sizes, and with some strong advice on "when to plug in the numbers."

We often *idealize* or approximate situations or objects when building models so that we can analyze the behavior of interest to us. We perform two kinds of idealizations—the first physical, the second mathematical. One very common idealization is to say that something is small, such as the range of motion of the classical linear pendulum. To translate the verbal statement of a physical assumption into a mathematical idealization or model is to say that angles will be small—that is,

$$\sin\theta \cong \theta, \quad \cos\theta \cong 1 \tag{1.39}$$

for some range of values $\theta \leq \theta_{pres}$ that is prescribed to be acceptable.

A simple mathematical approach to approximating the sinusoids in Equation (1.39) might be to expand either of them in traditional Taylor series; that is, we can represent the function $f(x)$ in a series about the point $x = a$ as

$$f(x) = f(a) + f'(a)(x-a) + \frac{f''(a)}{2!}(x-a)^2 + \cdots + \frac{f^{(n)}(a)}{n!}(x-a)^n + R_{n+1} \tag{1.40}$$

where the remainder term (which can be cast in several forms) is here

$$R_{n+1}(x) = \frac{f^{(n+1)}(\xi)}{(n+1)!}(x-a)^{n+1} \tag{1.41}$$

The derivative in Equation (1.41) is evaluated at a "suitably chosen" point ξ somewhere in the interval between a and x. Even though the precise location of ξ is not known, the remainder formula can be used to estimate the error made if a Taylor formula to order n is applied. For the kinds of approximations we generally make, however, we do not need to worry about this particular calculation. The reason is that we are typically looking for the smallest *nontrivial* or nonvanishing term in the series when expanded about a known reference point.

Equation (1.39) represents just such a case where we must be careful in translating assumptions, especially with respect to appropriate reference points. The approximation $\sin\theta \cong \theta$ is commonly for small angles of θ—although we will shortly show a case where that assumption would fail—but the second part of Equation (1.39) requires further thought about the meaning of small relative to a particular benchmark. In many physical models, the cosinusoid is linked to or compared with unity. For example, the height that a pendulum swings above its datum can be written as

$$y = l(1-\cos\theta) \neq l(1-1)$$
$$\cong l(1-(1-\theta^2/2!)) = l\theta^2/2 \tag{1.42}$$

from which we see that our approximation for the cosinusoid would have to be different from what we first indicated (in Equation 1.39) because its value is being compared to unity, not to zero. For example, had we been sloppy in translating this assumption when evaluating the pendulum's potential energy, we would have wrongly found that potential energy to be zero.

This kind of behavior is ubiquitous in engineering and physics, perhaps because much of the same mathematics is fundamental to modeling in those fields. For example, catenaries are found in bridges and in measuring tapes. The sag of such cables hanging under their own weight, h, is given in terms of a catenary constant c and the span l as

$$h = c\left(\cosh\frac{l}{2c} - 1\right) \tag{1.43}$$

It is clear that c has the same dimensions of length as do h and l. Now we substitute a Taylor expansion for the hyperbolic cosine to find the sag:

$$\frac{h}{l} = \frac{c}{l}\left(\cosh\frac{l}{2c} - 1\right) = \frac{c}{l}\left(1 + \frac{1}{2!}\frac{l^2}{4c^2} + \frac{1}{4!}\frac{l^4}{16c^4} + \frac{1}{6!}\frac{l^6}{64c^6} + \cdots - 1\right) \tag{1.44}$$
$$= \frac{1}{2!}\frac{l}{4c} + \frac{1}{4!}\frac{l^3}{16c^3} + \frac{1}{6!}\frac{l^5}{64c^5} + \cdots$$

For a tightly stretched string, the sag is very small compared to its length: $h/l \ll 1$. This suggests that the ratio $l/2c$ is also quite small compared to 1 because the one-term approximation of Equation (1.43) that follows from the last of Equation (1.44) is

$$\frac{h}{l} \cong \frac{l}{8c} \tag{1.45}$$

Equation (1.45) confirms the suggestion that large values of the dimensionless catenary parameter, $2c/l$, correspond to small values of the dimensionless sag, h/l, because this result can be arranged as

$$\frac{2c}{l} = \frac{l}{4h} \gg 1 \tag{1.46}$$

Further, had we approximated the hyperbolic cosine for small values of $l/2c$ independently of Equations (1.43) and (1.44), we would have calculated that

$$\frac{c}{l}\cosh\frac{l}{2c} \cong \frac{c}{l}\left(1 + \frac{1}{2!}\frac{l^2}{4c^2} + \frac{1}{4!}\frac{l^4}{16c^4} + \frac{1}{6!}\frac{l^6}{64c^6}\right) \cong \frac{c}{l} \tag{1.47}$$

We would then have found, quite mistakenly, that the sag was identically zero because we had used an inadequate approximation!

To illustrate this point still further, we now anticipate the estimation of arch behavior described in Chapter 4. As we will see, we will have to approximate terms such as the following for small values of the (dimensionless!) angle α:

$$f(\alpha) = 2\alpha + \alpha\cos 2\alpha - (3/2)\sin 2\alpha \tag{1.48}$$

Expressed in terms of standard Taylor expansions of the trigonometric functions, Equation (1.41) becomes

$$f(\alpha) \cong 2\alpha + \alpha\left(1 - \frac{(2\alpha)^2}{2!} + \frac{(2\alpha)^4}{4!}\right) - \frac{3}{2}\left(2\alpha - \frac{(2\alpha)^3}{3!} + \frac{(2\alpha)^5}{5!}\right) \tag{1.49a}$$

which, after regrouping by powers of α, appears as

$$f(\alpha) \cong \left(2\alpha + \alpha - \frac{3}{2}(2\alpha)\right) - \left(\alpha\frac{(2\alpha)^2}{2!} - \frac{3}{2}\frac{(2\alpha)^3}{3!}\right) + \left(\alpha\frac{(2\alpha)^4}{4!} - \frac{3}{2}\frac{(2\alpha)^5}{5!}\right) \cong \frac{4\alpha^5}{15} \tag{1.49b}$$

Thus, we had to keep terms out to $O(\alpha^5)$ in order to find a nontrivial result!

Now, all of the mathematical models of structural behavior that we present in this book are *linear*. This is because we *assume* that all of our structures can be modeled as having both a linear stress–strain law (cf. Chapter 2) and that the strains that we are analyzing are all so small that we can justifiably ignore the geometric nonlinearities that are part of more precise models of the mechanical behavior of solids. Thus, as a practical matter, all of the material we will present has already passed the "small strain" test of

$$\varepsilon\left(1+\frac{1}{2}\varepsilon\right) \cong \varepsilon \tag{1.50}$$

We show Equation (1.50) in part as a reminder that when we say the (dimensionless!) strain ε is small, we must say in relation to what, and here $\varepsilon \ll 1$. We also show this result to confirm our fundamental assumption—namely, that *all of the strains we will compute are very, very small when compared to unity.* This will also have implications for the relative magnitudes of particular structural deflections.

The fact that we can make the kinds of judgments we have described so far in this section also reflects a habit of thought, a mind-set with which solutions to problems of all sorts can be sought—and it is a mind-set that we will adopt for all of the work we do in this book and that we urge you to adopt for solving *all* engineering problems, not just structures problems. The fundamental idea is that when we make a calculation, we will proceed as far as possible with our *formulas left in symbolic terms.* That is, we will substitute numbers only as the *final* step in performing calculations. We do this because a symbolic result

- Enables checking the dimensions of a result, which is an essential part of model verification
- Makes it easier to relate a mathematical model back to its physical idealization and to reason about both
- Makes it easier to use a consistent set of units for a problem and to check the order of magnitude of the answer
- Leaves open the door to reusing a model because it is far more general in symbolic terms than it is when written in terms of problem-specific parameter values

The Assumption of Linear Behavior

The assumption of linear behavior is one of the most important concepts in mathematical modeling. Models of devices or systems are said to be *linear* when their basic equations—whether algebraic, differential, or integral— are such that the magnitude of their response is *linearly* or *directly proportional* to the magnitude of the input that drives them. Recall, for example, the Maxwell model of a viscoelastic material discussed just now. Even when devices like the classic pendulum are more fully described by nonlinear models, their behavior can often be approximated by linearized or perturbed models, in which cases the mathematics of linear systems can be successfully applied.

Linearity is applied during the modeling of the behavior of a device or system that is forced or pushed by a complex set of inputs or excitations. The response of that device or system to the sum of the individual inputs is obtained by adding or *superposing* the separate responses of the system to each individual input. This important result is called the *principle of superposition*. Engineers use this principle to predict the response of a system to a complicated input by decomposing or breaking down that input into a set of simpler inputs that produce known system responses or behaviors.

Linearity and Geometric Scaling

The geometric scaling arguments discussed previously can also be used to demonstrate some ideas about linearity in the context of *geometrically similar* objects—that is, objects whose basic geometry is essentially the same. Consider two pairs of drinking glasses: One pair is right circular cylinders of radius r; the second pair is right circular inverted cones having a common semivertex angle α. If the first pair is filled to heights h_1 and h_2, respectively, the total fluid volume in the two glasses is

$$V_{cy} = \pi r^2 h_1 + \pi r^2 h_2 = \pi r^2 (h_1 + h_2) \tag{1.51}$$

Equation (1.51) demonstrates that the volume is *linearly proportional* to the height of the fluid in the two cylindrical glasses. Further, since the total volume can be obtained by adding or *superposing* the two heights, the volume V_{cy} is a *linear function* of the height h. Note, however, that the volume is *not* a linear function of the radius, r.

In the two conical glasses, the radii vary with height. In fact, the volume, V_{co}, of a cone with semivertex angle, α, filled to height, h, is

$$V_{co} = \frac{\pi \tan^2 \alpha}{3} h^3 \tag{1.52}$$

Hence, the total volume of fluid in the two conical glasses is

$$V_{co} = \frac{\pi \tan^2 \alpha}{3} \left(h_1'^3 + h_2'^3 \right) \neq \frac{\pi \tan^2 \alpha}{3} (h_1' + h_2')^3 \tag{1.53}$$

That is, the relationship between volume and height is nonlinear for the conical glasses; the total volume cannot be calculated just by superposing the two fluid heights, h_1' and h_2'. Note that this result, while to a simple, even obvious case, is emblematic of what happens to superposition when a linearized model is replaced by its (originating) nonlinear version.

Conservation and Balance Principles

The development of mathematical models often starts with statements that indicate that some property of an object or system is being conserved. For example, the motion of a body moving on an ideal, frictionless path could be analyzed by noting that its *energy is conserved*. Sometimes, as when modeling the population of an animal colony or the volume of a river flow, *quantities that cross a defined boundary* (whether individual animals or water volumes) *must be balanced*. Such *balance* or *conservation principles* are applied to assess the effect of maintaining or conserving levels of important physical properties. Conservation and balance equations are related—in fact, conservation laws are special cases of balance laws.

The mathematics of balance and conservation laws are straightforward at this level of abstraction. Denoting the physical property being monitored as $Q(t)$ and the independent variable time as t, a balance law for the *temporal* or time rate of change of that property within the system boundary depicted in Figure 1.6 can be written as

$$\frac{dQ(t)}{dt} = q_{in}(t) + g(t) - q_{out}(t) - c(t) \tag{1.54}$$

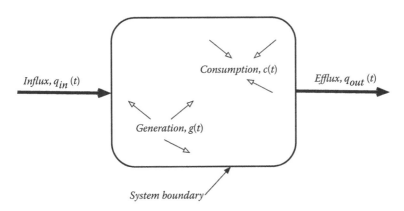

FIGURE 1.6
A system boundary surrounding the object or system being modeled. The influx $q_{in}(t)$, efflux $q_{out}(t)$, generation $g(t)$, and consumption $c(t)$ affect the rate at which the property of interest, $Q(t)$, accumulates within the boundary.

where $q_{in}(t)$ and $q_{out}(t)$ represent the flow rates of $Q(t)$ into (the *influx*) and out of (the *efflux*) the system boundary, $g(t)$ is the rate at which Q is generated within the boundary, and $c(t)$ is the rate at which Q is consumed within that boundary. Note that Equation (1.54) is also called a *rate equation* because each term has both the meaning and dimensions of the rate of change with time of the quantity $Q(t)$. In a mechanics context, the momentum $\mathbf{p} \equiv m\mathbf{v}$ is the physical property of interest, and the resulting balance law of interest is the *balance of forces* or *statement of equilibrium*:

$$\frac{d\mathbf{p}(t)}{dt} = \sum \mathbf{F}(t) \tag{1.55}$$

In those cases where there is no generation and no consumption within the system boundary (i.e., when $g = c = 0$), the balance law in Equation (1.54) becomes a *conservation law*:

$$\frac{dQ(t)}{dt} = q_{in}(t) - q_{out}(t) \tag{1.56}$$

Here, then, the rate at which $Q(t)$ accumulates within the boundary is equal to the difference between the influx, $q_{in}(t)$, and the efflux, $q_{out}(t)$. And in the mechanics domain, the resulting conservation law of interest is the *conservation of energy*:

$$\frac{d\mathbf{p}(t)}{dt} = 0 \tag{1.57}$$

Conclusions

This chapter has provided a very brief summary of the most basic foundations of mathematical modeling. In this context, the discussion began with a statement of principles under which the activity of mathematical modeling could be properly performed. This was followed by a discussion of basic foundational matters, including dimensional homogeneity and dimensional analysis, abstraction and scaling, balance and conservation laws, and an introduction to the role of linearity. It is important to note that this overview emphasized brevity, dictated by chapter length limitations, so it will hopefully serve as a stimulant to the reader's appetite for further reading and application of these basic ideas and methods.

Bibliography

Alexander, R. M. 1971. *Size and shape*. London: Edward Arnold.

Carson, E., and C. Cobelli (eds.). 2001. *Modeling methodology for physiology and medicine*. San Diego: Academic Press.

Cha, P. D., J. J. Rosenberg, and C. L. Dym. 2000. *Fundamentals of modeling and analyzing engineering systems*. New York: Cambridge University Press.

Dym, C. L. 2004. *Principles of mathematical modeling*, 2nd ed. Orlando: Elsevier Academic Press.

Gould, S. J. 1975. Size and shape. *Harvard Magazine* 78 (2): 43–50.

Heyman, J. 1995. *The stone skeleton*. Cambridge, England: Cambridge University Press.

Kemeny, K. 1959. *A philosopher looks at science*. New York: Van Nostrand-Reinhold.

Kline, S. J. 1965. *Similitude and approximation theory*. New York: McGraw-Hill.

Langhaar, H. L. 1951. *Dimensional analysis and theory of models*. New York: John Wiley & Sons.

Smith, J. M. 1968. *Mathematical ideas in biology*. London: Cambridge University Press.

Taylor, E. S. 1974. *Dimensional analysis for engineers*. London: Oxford.

Thompson, D'A. W. 1969. *On growth and form*. London: Cambridge University Press. (Abridged edition, J. T. Bonner, ed.)

Tufte, E. R. 1983. *The visual display of quantitative information*. Cheshire, CT: Graphics Press.

———. 1990. *Envisioning information*. Cheshire, CT: Graphics Press.

Problems

1.1 Use the basic method of dimensional analysis to determine the frequency ω (rad/s) of a beam given its Young's modulus E, shear modulus G, length L, second area moment I, cross-sectional area A, and mass density ρ.

1.2 Use the Buckingham pi theorem to determine the frequency ω (rad/s) of a beam given its Young's modulus E, shear modulus G, length L, second area moment I, cross-sectional area A, and mass density ρ.

1.3 Use the basic method of dimensional analysis to determine the frequency f (Hz) of a stretched membrane given a tensile force F per unit circumference, radius r, and thickness and mass density ρ.

1.4 Use the Buckingham pi theorem to determine the frequency f (Hz) of a stretched membrane given a tensile force F per unit circumference, radius r, and thickness and mass density ρ.

1.5 Would the results obtained in Problems 1.3 and 1.4 be different if ω were being sought, rather than f? Explain your answer.

1.6 Use the basic method of dimensional analysis to determine the frequency ω (rad/s) of a cantilever (of given Young's modulus E, length L, second area moment I, and mass density ρ) that also supports a concentrated mass at a distance x_0 from the base $(0 \leq x_0 \leq L)$.

1.7   Use the Buckingham pi theorem to determine the frequency ω (rad/s) of a cantilever (of given Young's modulus E, length L, second area moment I, and mass density ρ) that also supports a concentrated mass at a distance x_0 from the base $(0 \leq x_0 \leq L)$.

1.8   Use the basic method of dimensional analysis to determine the buckling load P of a column of given Young's modulus E, length L, and second area moment I.

1.9   Use the Buckingham pi theorem to determine the buckling load P of a column of given Young's modulus E, length L, and second area moment I.

1.10 Find the equation for the straight line drawn in Figure 1.3. (*Hint:* Render the church length and nave height dimensionless by dividing each by 1 ft.)

1.11 Determine the error made in the approximations (a) $\sin \theta \cong \theta$ and (b) $\cos \theta \cong 1$ for $\theta = 0°, 5°, 10°, 15°, 20°, 25°, 30°$.

1.12 Determine the error made in the approximations (a) $\sin^2 \theta \cong \theta^2$ and (b) $\cos \theta \cong 1$ for $\theta = 0°, 5°, 10°, 15°, 20°, 25°, 30°$.

1.13 How large can ε be for $\varepsilon(1 + \varepsilon/2) \cong \varepsilon$ to be good within 5%? 10%? 15%? 20%?

2

Structural Models and Structural Modeling

Summary

This book is about the business of modeling structures and their behavior. Thus, this chapter is devoted to describing in greater detail what it means to talk about *structural models* in the context of the modeling process.

Bars, Beams, Arches, Frames, and Shells

There are two broad categories of structures: those whose principal direction or line of action is coincident with the loads to which they are subjected and those whose principal direction or action is perpendicular to the direction of the applied loads. Within these broad confines we can further subdivide the domain of structural elements according to the number of physical dimensions needed to account properly for the behavior of interest—a number that is at least one, often two, but certainly no more than three! Thus, structures are often viewed as one- or two dimensional, depending on the relative magnitudes of a structure's physical dimensions and the number of (independent) spatial variables in the equation(s) of equilibrium in the structure's mathematical model.

The directionality and dimensionality issues interact in our attempt to sort structures into neat categories. This is because it is easier to create simplified, *idealized* models of structural elements or types than it is to ensure that all structures act in just such simple ways. However, our idealized models are widely representative of basic structural forms, and they do provide a vocabulary for identifying, describing, and analyzing more complicated *structural forms* and thus more complex structures.

There are cases where one- and two-dimensional idealizations will not do. For example, stress analyses of holes and cracks requires more detailed analysis, even when they occur in the gusset plates of trusses made up of one-dimensional bars. Such analyses are, however, considered more as aspects of three-dimensional elasticity theory than of structural analysis, so they are not discussed here.

One-Dimensional Structural Elements

In the first category we discussed, in which the loads and their attendant principal structural responses have the same directions, we are usually talking about structural elements that can be modeled as *one-dimensional structures*. These structures act in pure tension, as ropes or some of the bars in a truss, or they act in pure compression, again in truss bars and in columns, although columns deserve a category of their own. This is because they exhibit unusual behavior because of limits on the loads they can carry that are dictated by nonlinear geometrical concerns.

In Figure 2.1 we show one-dimensional structural elements and their dominant or major stresses. For this discussion, we will denote such dominant stresses by σ_n for normal stresses and by τ_s for shear stresses. We also note that each of these elements is far longer than it is deep or wide. Stated otherwise, the depth h and the width b of each of these elements are small compared to its length L, and "small" typically means less than 1/10.

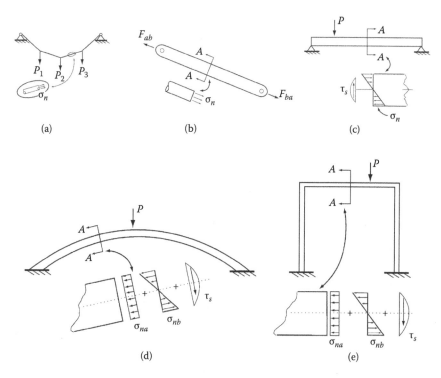

FIGURE 2.1
One-dimensional structural elements and dominant stresses: (a) cable; (b) bar; (c) beam; (d) arch; (e) frame. (Dym, C. L. 1997. *Structural Modeling and Analysis.* New York: Cambridge University Press. By permission of Cambridge University Press.)

We begin with the cables of Figure 2.1(a) wherein the loads are often directed either perpendicular to or at other nonzero angles with the cable, and the cable itself generally has "kinks" at points where the concentrated loads are placed. (Where the load is aligned with the cable, we have the simple case of a rope in tension.) The resulting stresses in a cable or rope are normal stresses that act along the cable's lines or segments, always in tension. Remember that you cannot push on a rope. Such cables are clearly mainstays of such fabled crossings as the San Luis Rey Bridge, and experience and intuition suggest that the cables of such bridges will need both horizontal and vertical support at the ends.

In Figure 2.1(b) we show bars, for which we assume that loads can be applied only at the endpoints, not along their lengths. Thus, each bar can be viewed as a single segment of the cables of Figure 2.1(a), although bars are more useful if we allow them to support both tensile and compressive normal stresses. This is because bars taken individually are of limited use, but, assembled into truss structures, they are very useful. Thus, bars and cables form a class of one-dimensional structures wherein the dominant stresses are normal stresses acting along the line(s) of the structure and the loads are applied, respectively, in directions that may depart significantly from those lines (for cables) and in arbitrary directions at the endpoints (for bars) at which they are joined to other bars to make up trusses.

The next one-dimensional structure we examine is the class of beams (cf. Figure 2.1c). Here the loads are almost invariably normal to the axis of the beam, and the beam supports these loads by a combination of normal bending stresses in the axial direction and transverse shear stresses perpendicular to the beam's axis. Recall that a beam cannot be supported solely by the transverse shear stresses because it is impossible to satisfy moment equilibrium without the bending stresses (cf. the beam section A–A of Figure 2.1c). Beams are useful because they transfer loads from locations along their span to the reactions that are typically located at the beam's ends. While doing this, they *bend* or deflect transversely. Thus, the beam is clearly the kind of structure intended to support loads perpendicular to its dominant geometrical line, and its visible behavior is its vertical deflection. However, unlike the cable of Figure 2.1(a), which arguably does the same, the beam has no horizontal reactions at its ends, only vertical reactions and, more often than not, moments.

The third one-dimensional structure, the arch, is shown in Figure 2.1(d). It provides an interesting contrast to the cable and the beam. The arch also carries loads normal to its basic direction; however, it is most efficient when it redirects its vertical loads to compressive normal stresses directed along the arch's axis and distributed uniformly over the arch's thickness. However, unlike the cable, which has only a pure normal stress across its cross section, a less than ideal arch uses both transverse shear and normal bending stresses—in addition to the dominant compressive normal stress—to carry its vertical loads to the reactions at the ends of the arch. We will illustrate some aspects of arch behavior in Chapter 6.

The final one-dimensional structure is the frame. In the same way that an arch has behavior in common with both the cable and the beam, the frame has its roots in beam behavior combined with bar-like aspects. Imagine, for example, that three slender beams are used as building blocks to erect the frame pictured in Figure 2.1(e). If the topmost frame element is something like a simple beam, it will need vertical supports at its ends to transmit its vertically applied load to the ground (through the frame's two vertical members). Thus, the vertical legs of the frames, though they might act largely as beams, must exhibit bar-like (or column-like) behavior to carry the vertical loads down to the ground.

Similarly, the horizontal bar atop the frame will combine both horizontal bar-like behavior with its own beam-like bending. Thus, the transverse or shear stresses of some frame members become the axial stresses of the members to which they are connected. And, in fact, frame members are generally modeled as beams that support uniform axial, bar-like stresses, in addition to their transverse shear and normal bending stresses. Arches are more typically used for some bridges and, occasionally, for some special effects in buildings, such as the proscenium arches of theaters and concert halls. Frame elements, on the other hand, are literally the building blocks of modern office and apartment building design.

Stress Resultants for One-Dimensional Structural Elements

We now provide an advance look at a descriptive device that is essential to modeling these elements. It will also significantly enhance our ability to model *supports* later.

We have chosen to display the responses of the structural elements just described in terms of the dominant stresses that occur in the cross sections of each particular structure. However, the analytical models of these structures are cast in terms of force stress resultants (also called force resultants) that represent integrated values of the dominant stresses, with the integrations being over the cross-sectional areas that are perpendicular to the axes of these elements.

For cables and bars, we simply integrate the uniform normal stress over the area and define an axial stress resultant $N(x)$ as (cf. Figure 2.2a)

$$N(x) \triangleq \iint_{A=hb} \sigma_n \, dz \, dy \tag{2.1}$$

Thus, $N(x)$ represents the *axial force resultant*, tensile or compressive, produced by the dominant normal stress across the bar or cable. As a matter of notation, for cables and ropes the notation $T(x)$ is often used, reflecting the fact that these structures are always in tension. Similarly, when bars are assembled in trusses, the force in each bar is typically denoted by F_{ij}, where

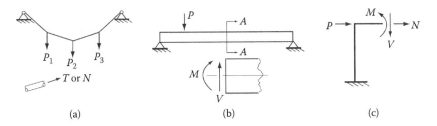

FIGURE 2.2
Stress resultants for (a) cables and bars; (b) beams; and (c) frames. (Dym, C. L. 1997. *Structural Modeling and Analysis.* New York: Cambridge University Press. By permission of Cambridge University Press.)

the subscripts stand for the numbers of the joints that the bar connects. These bar forces can be either tensile or compressive.

For the beam, the dominant stresses are the normal bending stress σ_n and the transverse shear stress τ_s. These are integrated differently because the bending and shear stresses are distributed differently over the cross section. The normal bending stress is itself distributed linearly through the thickness of the beam (it is constant across the width, as are all beam variables); however, its net average across the thickness is zero because of this linear distribution. On the other hand, the integral of the first moment of the bending stress about the centroid of the beam's cross section produces the *moment* or *bending stress resultant* $M(x)$, defined as (see Figure 2.2b)

$$M(x) \triangleq \int_h \sigma_n z b\, dz \qquad (2.2)$$

The transverse shear stress generally varies quadratically over a beam's cross section. However, we are less interested in the details of that shear stress distribution than in the resultant (or simple average), so we define the *shear force* or *shear stress resultant* $V(x)$ as (see Figure 2.2b)

$$V(x) \triangleq \iint_{A=hb} \tau_s\, dz dy \qquad (2.3)$$

For frames and arches, we would expect that all three of the aforementioned force and moment resultants will be present in the most general case, for reasons we have already noted. However, while axial, shear, and bending resultants are invariably present in frames to satisfy static equilibrium, their individual effects on the deformation of a frame will be substantially different. Similarly for arches, where we have noted that the ideal—or most efficient—arch has only a compressive normal stress distributed across its thickness, there will be both shear and bending effects. For the arch, however, these three resultants are often coupled in interesting ways by the very curvature of the arch, as we will see in Chapter 6. However, to be most general or inclusive, it is best to assume that both frame and arch cross sections are acted upon by the complete set of force and moment resultants, per Figure 2.2(c).

Two-Dimensional Structural Elements

There are several important two-dimensional or *surface* structures, both planar and curved. The first such surface structure is a two-dimensional counterpart of the cable, called the *membrane*. A membrane supports vertical loads through purely tensile normal stresses (cf. Figure 2.3a), except that the two-dimensional nature of the membrane renders it unable to kink like a cable. Perhaps the most common form of a membrane is the trampoline, but membranes are often used as roofs or domes over stadia and are partially supported by the air pressure contained by the roof.

In fact, membranes are not often used in buildings because they require enormous tensile stresses and forces in order to maintain a virtually flat surface under loads applied perpendicular to that surface. This in turn means that there needs to be some massive structure erected to support those in-plane stresses and forces. In fact, there are parallels between the membrane support requirements and the comparable situations for cables and arches. It is no surprise, for example, that arches are often contained between immobile riverbanks or rock faces or that they are built upon massive haunches— all to support the large compressive stresses and forces that result from the redirection of the loads carried by the arch.

The next surface structure of interest is the *plate,* shown in Figure 2.3(b), which is a two-dimensional analog of the beam. The basic behavior is just

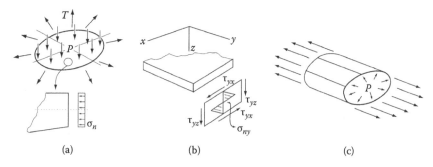

(a)   (b)   (c)

FIGURE 2.3
Two-dimensional structural elements and dominant stresses: (a) membrane; (b) plate; and (c) shell. (Dym, C. L. 1997. *Structural Modeling and Analysis.* New York: Cambridge University Press. By permission of Cambridge University Press.)

like that of beams in that transversely applied vertical loads are converted into normal bending and transverse shear stresses, although the extensions from beam theory are somewhat complicated by the nature of the two-dimensional modeling needed to describe the bending of a plate structure. Plates are used in a variety of structures, mostly to model aspects of the *cladding* or covering of a structure, including everything from roofs to windows, and also to model coverings over holes in the ground, whether manholes or construction trenches.

In Figure 2.3(c) we show *shell* structures (or simply *shells*), which reflect combined behaviors that arise because they are *curved* surface structures. In fact, shells are often described as curved plates, but their behavior is more complicated because a shell's curvature couples the in-plane normal stresses to the transverse shear stresses. Further, the in-plane normal stresses have two components: one uniform, like the axial stresses in bars and arches, and a second that varies linearly through the thickness, like the bending stresses in beams and plates. Again, shell models are more complicated because of the curvature effects, and their attendant behavior is also more complicated—and often more interesting. Shells are used in storage structures, such as grain silos and oil tanks, and in dramatic roof structures, such as the hyberbolic paraboloids and other curved surfaces that often provide graceful roofs for modern houses of worship. Shell structures are also evident in aircraft and spacecraft, and in objects as mundane as trash cans.

Modeling Structural Supports

While we have made a few comments about the reactions needed to support some of the one- and two-dimensional structures we described, we have not said nearly enough. As this is obviously a very important topic, we turn to it now: How do we support a structure? And, equally relevant for our purposes, how do we model such supports?

It is clear that to support a structure means that we want to keep it from moving or going anywhere and that we want to ensure that the loads applied to the structure are constructively carried down into a foundation attached to the ground that, for the most part, is considered the immovable object on which our structure can depend for support. The reason that the modeling aspect becomes important is that the preceding statement—although fairly abstract—is (generally) quite correct. We need to express the need for support reactions—or for permissible deflections or displacements—in the language used to describe the behavior of the structures of interest. This means that the kinds of reactions or supports are best described here in rather general terms, with appropriate details left for the development of the individual structural models.

There are two basic themes for describing support reactions. The first is that the number of generalized force reactions needed at each support is exactly equal to the number of independent resultants (force and moment) required to write a complete set of equilibrium equations for the structural model of interest. We say generalized force reactions to accommodate the fact that some of our force resultants are moments and thus, strictly speaking, are not forces.

The second theme is that each generalized reaction force has a generalized displacement *dual*—that is, a displacement quantity that can be prescribed *instead of* that force, although not in addition to the force. The displacement dual is that generalized displacement through which the corresponding reaction force would move to do work.

Thus, as a very simple example, consider the rope or cable shown in Figure 2.4(a). Inasmuch as this cable requires only a statement of equilibrium regarding its uniform axial stress (or its net axial force), we need to stipulate only one reaction force, or its dual, at each end. At the top end, since it represents the suspension point from which we are hanging the cable and its load, there is only one reaction force, and it must balance the tension in the cable. At the other end, we could require the force we want the cable to carry (say, a weight W) or we could require that the cable be stretched by a known amount (say, Δ). However, we cannot stipulate both W and Δ because these two quantities are duals of one another: Their product is directly proportional to the work done at the cable end, and we cannot independently prescribe both the force and the displacement needed to do work there. Stated in the context of a very simple yet familiar example, we cannot independently require specified values for a force applied to extend a linear spring and to the displacement that would result.

For a bar, whether alone or as a member in a truss, the situation is very much the same as that for the cable or rope. Although the components of forces applied to a bar at its ends may take on any direction, their net effect or resultant must be collinear with the bar, and the net deformation of a bar is simply the extension or compression of its length.

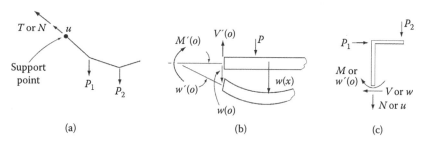

FIGURE 2.4
Force–displacement duals at supports of (a) cables or ropes; (b) beams; and (c) frames. (Dym, C. L. 1997. *Structural Modeling and Analysis.* New York: Cambridge University Press. By permission of Cambridge University Press.)

The beam is more complicated because two independent generalized forces, a moment and a shear, are needed to describe how that beam supports it load. Thus, at each beam support $x = x^*$, there are two choices to be made (cf. Figure 2.4b): prescribe either a shear force $V(x)$ or its corresponding dual, the transverse or bending deflection $w(x)$ of the beam, *and* prescribe either a moment $M(x)$ or its corresponding dual, the slope of the bending deflection of the beam. Thus, the beam support alternatives can be stated as

$$\text{At } x = x^* \text{ prescribe either } V(x) \text{ or } w(x) \tag{2.4a}$$

and

$$\text{At } x = x^*, \text{ prescribe either } M(x) \text{ or } w'(x) \tag{2.4b}$$

For a simple beam with supports at both ends—say, $x = 0$ and $x = L$—there are two possibilities at each end and thus four possible permutations of ways we can support a two-ended beam. Some of these are sketched in Figure 2.5, alongside drawings of how such kinds of supports are implemented in the real world of structural engineering. The two basic support cases are (1) the *simple support* or *pinned* case, which requires that both the bending displacement $w(x)$ and the moment $M(x)$ vanish, and (2) the *clamped* or *fixed* case, which requires that both the bending displacement $w(x)$ and its slope $w'(x)$ vanish. There are several ways these supports are implemented, as shown in Figure 2.5, but perhaps the most interesting aspects are only implicit in these pictures.

For example, for a simple girder bridge, such as those used for highway and railroad bridges, or something that evidently functions just like a beam within a framed structure, we often see just the sort of pins and

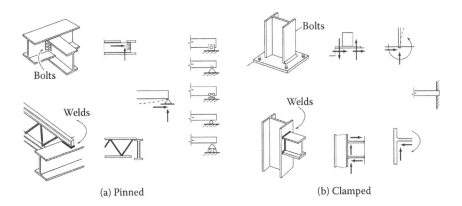

(a) Pinned   (b) Clamped

FIGURE 2.5
Classical supports of (a) pinned and (b) clamped beams when they stand alone or when they are attached to vertical members in frames.

girder connections shown in Figure 2.5. However, it is appropriate to wonder whether or not these boundary conditions are simply limited to the shear and moment choices given in Equations (2.4a) and (2.4b), or whether or not we need to account for any horizontal motion and its axial force dual.

For bridges, it may be reasonable to ignore the horizontal force–displacement duality because expansion joints are purposely built in to ensure a freedom of horizontal motion due to temperature expansion. For framed structures, if a beam is situated atop two vertical legs, as with the frame of Figure 2.1(e), then we clearly have to ensure that any horizontal motion of the beam is compatible with any transverse motion of the legs, and we have to ensure the equilibrium considerations of the beam's axial load being transmitted between the two column legs.

But the most interesting aspect is that the beam–column frame connections shown in Figure 2.5 cannot be as black and white as the choices offered by Equations (2.4a) and (2.4b) because, in general, we cannot literally expect deflections or slopes to be zero when beams are attached to columns that are not themselves perfectly rigid. That is, there is a fundamental modeling assumption contained both within our discussion of the cable of Figure 2.4(a) and for the elementary beam supports: Namely, there is a device that will provide an absolutely immovable support against the bending displacement or the bending slope (or both) of a beam. This is a good model for a variety of practical circumstances, but it is not universally applicable. In particular, in the case of frame structures (cf. Figure 2.4c), the relative stiffnesses of beams and of the columns to which they are attached become very important in analyzing frame behavior.

And, of course, we are doing our best to model the reality of supports as they would appear and function in actual construction. Thus, support conditions, both those discussed here and those applied in more complex structural calculations, are intended to reflect realistic support behavior whose effects we wish to take into account. So, for our physical and mathematical models, we should choose boundary and support conditions with an eye toward what is really out there in the world.

Indeterminacy, Redundancy, Load Paths, and Stability

One of the issues that derives from the nature of the support conditions is their *determinacy* or, more accurately, their *statical determinacy*. That is to say, given a structure and a set of loads that act on that structure, we have to answer a question: Can a structure's support reactions be calculated solely from the equations of static equilibrium? It turns out that sometimes they can, and sometimes they cannot. The basis of the answer is simple enough. If there are enough independent equations, we can calculate the unknown reaction forces. If there are not enough equations and if there are more unknown reaction forces than available equilibrium equations, then we cannot. How do we know? And does it matter much whether a structure is classified as statically determinate or as *statically indeterminate?*

Counting Degrees of Indeterminacy

First, on how we know whether a structure is determinate, it is a relatively simple matter to count up the unknowns and compare that number with the number of available equations. We could do this in the abstract, for three-dimensional elastic solids, and we could do it for individual cases or models. For one-dimensional bars taken alone, for example, we have only one equilibrium equation, so it is easy enough to inspect a particular bar's configuration and assess its determinacy by counting the number of supports at which a reaction force is required (as opposed to bar ends where displacement is specified or permitted).

When bars are assembled into trusses, the situation is more complicated. For *planar trusses*—that is, trusses that act in a plane—we can write three equilibrium equations (i.e., $\Sigma F_x = 0$, $\Sigma F_y = 0$, $\Sigma M_z = 0$) for the truss in its entirety. However, we can also write two equations (i.e., $\Sigma F_x = 0$, $\Sigma F_y = 0$) for each truss joint, and if we do that for every joint in the truss, we automatically include the three equations taken for the truss in its entirety. Thus, the three truss equations are subsumed in the complete set of joint equations because, at the end of each bar, the resultant forces are colinear with the bar, so $\Sigma M_z = 0$ for the bar. Thus, there are $2j$ independent equations for a truss with j joints. On the other hand, in terms of unknowns, we have r reaction forces and b bar forces—that is, the resultant tensile or compressive forces that occur in each bar. The test for the determinacy of a planar truss is, then,

$$r + b = 2j \Rightarrow \text{determinacy} \tag{2.5a}$$

$$r + b > 2j \Rightarrow \text{indeterminacy} \tag{2.5b}$$

The case $r + b < 2j$ involves stability issues that will be discussed later.

The kind of calculation we have done can clearly be extended to other models—for example, space (three-dimensional) trusses. For our purposes, the most important "other case" is the set of problems governed by models of beam behavior. Here, for planar beams, we can recognize that, once again, there are three equilibrium equations (i.e., $\Sigma F_x = 0$, $\Sigma F_y = 0$, $\Sigma M_z = 0$) and that the number of potential force unknowns can be counted by simply looking at the support reactions. We must be a bit careful, however, to recognize that there is no net deformation (and hence no net axial force resultant) in elementary beam theory, so the horizontal equilibrium equation (i.e., $\Sigma F_x = 0$) is typically useful only for calculating the single horizontal reaction required to keep a beam from rolling away on a horizontal plane. On the other hand, we must count the horizontal equilibrium equation as having greater import when we are analyzing frames because (1) the horizontal reactions of a frame have more physical meaning and impact, and (2) the axial forces on the horizontal bars of a frame are the lateral loads that bend the frame's vertical legs.

Coming back to beams, now, at a pinned or simply supported support, we need to account only for the vertical reaction (in shear), while at a clamped support we must account for both the shear and the moment. Thus, a beam clamped at both ends has four unknowns, and discounting the horizontal, we have only two equations (i.e., $\Sigma F_y = 0$, $\Sigma M_z = 0$), so the beam is indeterminate to the second degree.

Now, why should there be more reactions than we can calculate with the equations of static equilibrium? Is it intentional? What do we gain by incorporating "extra" reaction forces? After all, if we have just enough unknowns that we can calculate them all with the equations of equilibrium, do we not have a structure that is in equilibrium and will not go anywhere? That is, if all of the equations of static equilibrium are properly satisfied, are we not told by Newton's laws that we have a structure that is (statically) stable? Why do we need these extra, redundant forces? Are we simply interested in making it harder to analyze indeterminate structures (a view often held by students in structures courses)?

We will have occasion to visit the issue of degree of indeterminacy for beam and arch problems, and we will then discuss some techniques for handling such problems. For now it is important to recognize that there are many problems for which the external static equations of equilibrium are insufficient for analyzing a structure, and that many structures are intentionally designed to be statically indeterminate. So, while we defer discussions of techniques for analyzing indeterminate structures, it is a good time to discuss the reasons why structures are designed to be that way.

Indeterminacy and Redundancy Matter

The issue of indeterminacy matters. We have already noted that any reaction or supporting forces beyond the minimum number needed to satisfy the equations of static equilibrium are extra or, in an equivalent word, redundant. Thus, we will use the term *redundant* to refer to each and all of the forces beyond the minimum number required by static equilibrium.

Why are redundants designed into structures? What purposes do redundants serve? The answer has two parts that are, briefly, as follows. First, indeterminate structures are generally stiffer than their determinate counterparts, so the generalized forces (and thus the stresses) and the deflections they experience will be less than the corresponding results for their determinate counterparts. Second, the existence of redundant reactions means that the failure of some of the supports could still leave a viable, determinate structure that remains in equilibrium and, thus, is still stable.

On the first point, consider the maximum bending deflections and stresses for fixed-ended and simply supported beams of length L that support a uniform load-per-unit length q_0. We recall that the (dimensionless) maximum displacement $EIw_{fixed}(L/2)/q_0L^4$ is reduced from 5/384 to 1/384—by 80%—and the (dimensionless) maximum moment $M(L/2)/q_0L^2$ from 3/24 to 1/24—by 67%—when we move from a fixed-ended beam to the much more flexible (i.e., much less stiff) simple beam. Thus, we increase the stiffness substantially by moving from a determinate simply supported beam to an indeterminate fixed-ended beam. It is virtually impossible to quantify such stiffness gains in general in a meaningful way. However, it certainly is intuitively evident that by increasing the number of redundants we are simultaneously restricting a structure's ability to deflect and deform, so we are stiffening that structure. For the uniformly loaded beams just considered, for example, it is clear that clamping the supports restricts the beam's flexibility in comparison with that of the simply supported beam. Thus, we effectively increase a beam's stiffness by clamping its ends. It should therefore not be in the least surprising that our vast experience with structures confirms this idea.

On the second point, consider the two beams shown in Figure 2.6(a, b), which we might imagine as bridges over a simple gap. Imagine that the right support of each washes away, resulting in the situations shown in Figure 2.6(c, d). The first beam, originally simply supported (and thus statically determinate) becomes a *mechanism* with the loss of its right-hand support: What was once an elastic beam has now become the equivalent of a *rigid bar* whose only support is the pin at the left support, which is unable to supply sufficient constraint to keep the structure in a stable state. The rigid bar will simply rotate freely about its remaining support.

But as shown in Figure 2.6(d), the second bridge, originally fixed at both ends, now resembles a diving board: It is now a cantilever beam that carries a uniform load, so it remains a stable structure. It will not be as stiff as the original fixed-fixed beam, but it does support the original load. We also see that an alternate *load path* has been provided to carry the applied loads through to the bridge's remaining foundation. The redundant

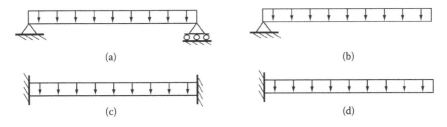

FIGURE 2.6
Loss of supports of two beams as bridges. (Dym, C. L. 1997. *Structural Modeling and Analysis.* New York: Cambridge University Press. By permission of Cambridge University Press.)

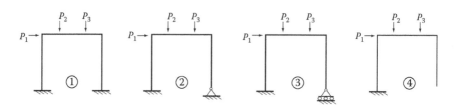

FIGURE 2.7
Alternate supports and load paths in some simple frames. (Dym, C. L. 1997. *Structural Modeling and Analysis.* New York: Cambridge University Press. By permission of Cambridge University Press.)

reactions thus provide *redundancy*—that is, safety for the bridge provided by a redundant load path.

This is also the case for the sequence of frames shown in Figure 2.7. Each is stable, and each progresses from a highly redundant structure (indeterminate to the third degree) to one that is clearly determinate. It is as if the right support had slowly weakened and eventually given way entirely. What remains is still a stable structure; it is just considerably more flexible, or less stiff, than the original structure, although it does support the same loads. The set of reactions on each frame will be different, with the changes occurring as another redundant reaction is removed. This progression of structures shows that we are providing alternate load paths by which the applied loads are carried down to the frame's foundation. Thus, as with the simple-beam bridge, we are providing additional safety for the frame by providing redundant load paths.

Such redundancy shows up in other ways. For example, it might seem impossible to guard against leaks due to cracks or ruptures in the skins of storage tanks designed to hold liquids or gases, just as it is impossible to contain within a burst balloon the pressurized gas that inflated it. However, it is now routine for oil to be transported by sea in tankers that are *double hulled* so that, should a tanker run aground or hit a reef, there is a redundant layer of protection that guards against oil spills. The environmental and other costs attached to major oil spills are so large that the extra expense of designing and building oil tankers to be redundant is considered a sound business decision.

Perhaps it is a commentary on our times, but similar issues are now being routinely incorporated in building design. One tragic spur to such considerations was the bombing of the Alfred P. Murrah Federal Building in Oklahoma City on April 19, 1995. It has been estimated that 80% of the 168 deaths that day resulted from structure failure, rather than as a direct result of the blast. Thus, the issue of extending building criteria to include structural measures against progressive collapse is a very pertinent topic, especially in terms of regulations for government office buildings, as well as for commercial buildings. The point is that structural protection against progressive collapse can be provided by incorporating designs of alternate load paths to minimize the effects of losing critical structural elements, no matter what causes these critical elements to fail.

There are many considerations that enter into building design. For example, design considerations for a public building such as the Murrah Building include cost, aesthetics, public access, and the very appearance of public accessibility for a government office building in a democracy. And there are certainly other safety measures that can be, and are being, taken into account. Our point is simply that there are good reasons to design and build indeterminate structures and that these are some of the challenges of structural engineering that we can at least appreciate at this (early) point.

Important Aspects of Structural Stability

It is interesting to enter into a discussion of stability immediately after our remarks on the role of redundancy in structural design. No doubt we all sense that we have some gut-level feeling or intuition of whether a structure is stable, or whether it might suddenly collapse at our feet. And certainly anyone could look at a picture of the Murrah Building after the explosion and wonder whether the remaining structure was itself stable—that is, whether it would stand for long on its own. In fact, there are two different technical meanings attached to the notion of stability, one of which is related to redundancy.

The meaning of stability that is related to redundancy is also partially related to the test for the determinacy of a planar truss that was given in Equations (2.5a) and (2.5b). In the notation of those equations, we now ask: What does happen when $r + b < 2j$? This question is the converse of the test for determinacy, and its answer is that in this case the total number of reaction forces and bar forces is too small to satisfy equilibrium at all of the joints.

Remember that the satisfaction of equilibrium in Newtonian mechanics implies that there is a sufficient number of *constraints* to keep the structure from moving. Further, these constraints are both *internal,* as manifested in bar forces, and *external,* as manifested in reaction forces. We show a classic instance of this in Figure 2.8(a), noting that this structure is underconstrained internally (and its attributes satisfy the given inequality because $r + b = 3 + 4 = 7 < 2j = 2(4) = 8$).

On the other hand, the truss in Figure 2.8(b) is both stable (remember that trusses are assemblies of triangles for good reason) and determinate, and for which $r + b = 3 + 13 = 16 = 2(8)$, which is twice the number of joints, so Equation (2.5a) is satisfied. It is also easily seen that if another bar is added to the truss in Figure 2.8(b)—say, another diagonal across one of the two center panels—the new truss would be internally indeterminate because while we can still calculate the three reactions, we do not generate any more equations by adding just that bar, so now we cannot calculate all of the bar forces (now numbering 14).

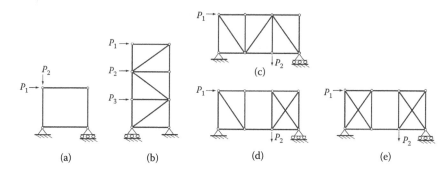

FIGURE 2.8
Internal determinacy and stability in trusses. (Dym, C. L. 1997. *Structural Modeling and Analysis.* New York: Cambridge University Press. By permission of Cambridge University Press.)

The relationship between determinacy and stability is sufficiently complex—especially for trusses—that simply applying formulas is not enough to determine determinacy and stability. Consider the set of trusses in Figure 2.8(c–e). The first two (c, d) have the same numbers of reactions, bars, and joints ($r + b = 3 + 13 = 16 = 2j = 2(8)$), yet the first of these trusses is determinate and stable, while the second is determinate and unstable. Further, we can make this truss both unstable and indeterminate by adding another bar (e). In both of the last two trusses, we have created a *mechanism* by virtue of making the center panel unable to remain rigid or, in another view, of making that panel (and thus the truss) unable to transmit a vertical load across that center panel. Thus, a structure can be unstable while being either determinate or indeterminate.

The idea of a mechanism derives from the fact that large rigid-body motions of some structural elements result from these hinges, and the elasticity of the members joined at the hinges becomes irrelevant as each member behaves as a rigid bar or link. These assemblies of links and hinges that permit such large rigid-body displacements of a structure, or pieces of a structure, are thus called mechanisms, no doubt with an eye toward "mechanical" mechanisms that are designed as ensembles of links and connectors such as hinges and springs.

Note that the mechanisms we have shown so far are instances of *internal instability* that result from insufficient internal constraint. It is also true that *external instability* can result if all of the supports of a truss are either *parallel* or *concurrent*. In fact, this is true for any kind of structure, whether it be a truss, a beam, a frame, an arch, or whatever. In these cases, the possibility of unstable structural behavior occurs because there is no external constraint against either a force directed at an angle to the parallel reactions or a net moment about the point of concurrency.

Internal mechanisms also appear in beams and in framed structures, typically because of the presence of *hinges*—that is, points of zero moment.

This is not to say that all hinges produce mechanisms, but rather that hinges are either a signpost of an unintended mechanism or a device used to allow or insert unstable behavior in a structure.

The other kind of stability worth noting is of a markedly different kind, perhaps most easily illustrated for the Euler buckling of a *column* for which *the onset of bending behavior* occurs at axial loads that are determined by solving an *eigenvalue problem* centered on the following differential equation:

$$EI\frac{d^4w(x)}{dx^4} + P\frac{d^2w(x)}{dx^2} = 0 \tag{2.6}$$

Now, it is important to recognize two facets of column behavior. The first is that the mathematics of column behavior are inherently nonlinear. That is, even though Equation (2.6) is a linear differential equation, the column's loss of stability is determined by a critical value of one of the coefficients in this equation, the value of the load $P = P_{crit}$. Thus, we cannot apply superposition as we typically would wherein we expect that the response of a structure (or any linear system) to two loads (applied as forcing functions on the right-hand sides of the equilibrium equation or equations) can be written as the sum of the responses to each load applied independently.

The second facet of column behavior is illustrated by the fact that the column shows no bending response $w(x) = 0$ for loads smaller than the critical load $(P < P_{crit})$ and that we cannot determine the magnitude of the column's bending deflection when the critical load is reached $(P = P_{crit})$. This is one reflection of the truly nonlinear nature of the column and similar structural stability problems, which is only partially reflected in the *linearized version* of the problem given in Equation (2.6).

The practical import of this is, in brief, as follows. The critical load of a column can be regarded as an upper limit of the load that can be supported by the column before unwanted transverse or bending deflections (perpendicular to the column's axis) can occur. In actual structural practice, frame members often serve as *beam–columns,* carrying both transverse and axial loads, each of which produces bending deflections. In this case the kinds of elementary models used to analyze framed structures have to be extended. However, we will be content in our discussion to observe that columns are important structural elements that require further study and careful design.

Modeling Structural Loading

No discussion of modeling structures can be completed without discussing both the loads that are applied to structures and the materials of which they are made. The sources from which structural loads derive include the dead loads of the structure's innate self-weight, live loads resulting from the traffic appropriate to the structure's use, and environmental loads that include the effects of wind, earthquakes, and rain and snow. However, we have said nothing about how those loads are actually modeled in the context of structural analysis and design. That is, once we have chosen a model for the structure, whether it be a truss, a beam, an arch, a bridge, or a complex framed structure, how are the loads actually applied?

There are two parts to the answer to this question. The first is concerned with how a set of loads is applied—whether as a set of forces or through a prescribed movement, such as are used in seismic analysis—and about how these loads are distributed in space and time. Wind loads, for example, can be felt as short-term gusts or as long-term steady winds, while seismic movements tend to last only briefly. Further, the aerodynamic forces resulting from wind blowing across a slender suspension bridge are going to produce a markedly different effect than the spatial distribution of wind pressure over a very tall and slender building; the latter will be strongly influenced as well by whether the building is surrounded by other tall structures or is more of a stand-alone kind of skyscraper.

The effects of some of these types of loads are so complex that part of structural modeling includes expensive physical modeling and testing. For example, actual scale models of skyscraper designs are now routinely tested in large wind tunnels, with appropriate inclusion of models of significant neighboring structures. On the other hand, some of the distributions can be successfully modeled in rough terms—for example, by postulating that wind pressure is distributed linearly over a building, from a zero value at its base to a maximum value at the roof.

The second part of the answer to the question about how loads are applied has to do with the connection between the model or idealization of a structure, on the one hand, and the kinds of loads on the other. The clearest example of this interaction is encapsulated in a simple planar truss, whether it is used as a frame for supporting a roof or as one side of a railway or highway bridge. Trusses, for example, can only be loaded at their joints or pins (because each of their bars is an axially loaded member). However, common sense tells us that roof trusses have to support loads spread out over the roof's surface and bridges have to support the traffic on the continuous railroad tracks or roadways that they are supporting. The answer here is a simple one—namely, that there are intervening structural elements that pick up the spatially distributed loads and "convert" them to the point loads of the trusses.

Now, complete load analyses are well beyond our focus on the estimation to structural behavior. We will generally use fundamental static models for our modeling, all of which should ring familiar: concentrated forces and moments, uniformly distributed line loads (per unit length of a one-dimensional structure), line loads with a triangular distribution, occasionally some other spatial polynomials, and, of course, loads that can be made up by superposing the fundamental loads.

Modeling Structural Materials

Like many of our other discussions, a complete discussion of structural materials is multidimensional. It clearly must include considerations of intrinsic material properties, such as the modulus of elasticity E, with dimensions of force per unit area, and the density ρ, having dimensions of mass per unit volume. However, knowledge of such basic properties is not a sufficient basis for choosing structural materials. Some of the general materials issues we need to understand include different aspects of *material behavior,* such as elastic limits, ultimate strengths, and crack initiation characteristics; how *combinations of material properties* compare; how materials are used in *varying configurations,* because their geometry is often very important; the influence of the *environment;* and, finally, the *cost,* including both unit material costs and costs that arise from different fabrication and construction requirements.

Wood, for example, is a structural material whose strength-to-weight ratio is high, but whose strength limits are below what would be needed for, say, use in high-rise building design. Further, wood's properties are nonuniform as they vary with grain orientation and with cell structure, facts well known to anyone who works with wood as a hobby. These two properties are also reflected in another potential limitation of wood construction—namely, that it splinters and cracks and fails dramatically when its strength is exceeded. This suggests that one of the material properties of interest should be its behavior during degradation or failure because graceful failure, say, of excessive deflection without rupture is preferable to a flat-out collapse or other catastrophic failure.

Thus, *ductile* materials (like steel, aluminum, and titanium) that continue to support significant loads while undergoing undesirably large deformations are often preferable to *brittle* materials (like concrete and stone) that fail by cracking. However, the failure considerations are also strongly affected by the structural configuration; for example, the cracking of a single bar in a wooden roof truss is a different matter than the failure of a high-strength cable on a long suspension bridge.

The choice of material behavior is also strongly influenced by the environmental circumstances and operating conditions in which the material is used (e.g., temperature) and by the geometrical configurations that are employed. In the first category we might need to recognize that steel can exhibit brittle behavior in very cold environments and can corrode due to salt water or acid rain, while for concrete structures we might well have to worry about the prevailing climate when the concrete is being poured and during its curing time. In the second category we note a venerable building material, reinforced concrete, in which two materials, steel and concrete, are combined in a particular configuration—slender steel reinforcing bars in the lower halves of thick concrete beams or floor slabs—to take good advantage of their individual behavioral properties, the ductility of steel in tension, and the effectiveness of concrete aggregate in compression.

One of the most interesting, and most challenging, materials issues is the relation between structural function, structural configuration, and structural material—for example, whether to design a bridge as a girder or an arch or a suspension bridge, and if an arch, whether it should be made of reinforced concrete or erected as a steel framework. Among the obvious variables that will influence these choices are the width of the gap that is being spanned, the aesthetic judgment of the owner or sponsor, and the availability of funding.

Clearly, there are a lot of materials issues—many more than we have discussed. Equally clearly, we will have to limit our materials modeling at this stage. The model we will use throughout is derived from a classic stress–strain curve, such as the one shown in Figure 2.9. In that picture we note various stages of behavior for a mildly ductile steel. Of particular interest is that, as exemplified by a standard uniaxial test, most of the stress that a piece of steel can carry is reached during that portion of the curve where the stress and strain are linearly related. Thus, as our one-dimensional model of material behavior, we take

$$\sigma = E\varepsilon \tag{2.7}$$

This model will serve us well for most one-dimensional applications. To include shear in an otherwise one-dimensional model or to model two- or three-dimensional structures, we generalize Equation (2.7) to *isotropic elastic materials*, whose material properties do not change with orientation and for which we need only account for Poisson's ratio, v. Most importantly, though, Equation (2.7) and its multidimensional counterparts reflect the idea that we

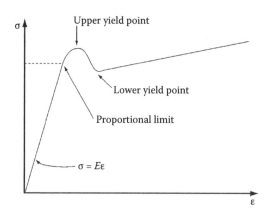

FIGURE 2.9
A stress–strain curve for mildly ductile steel. (Dym, C. L. 1997. *Structural Modeling and Analysis.* New York: Cambridge University Press. By permission of Cambridge University Press.)

will generally be dealing with linear models of material behavior, which is one of two reasons behind our ability to say that all of the models discussed in the sequel represent various instances of *linear springs*. That is, to paraphrase Shakespeare, "All the world's a linear spring; the only question is what is the stiffness k?"

Idealization and Discretization in Structural Modeling

Our final note on modeling is brief. In the previous section we talked explicitly of the idealization part of the modeling process, wherein we develop an abstraction or an approximation of something that we wish to analyze. We do this, of course, because we can deal with only so many variables at any given moment and, even more importantly, because the phenomenon or device we are modeling does not always require an infinite amount of detail to provide a useful understanding. Thus, we start modeling by seeking an *idealization*. For much of our work, the idealization will be expressed in differential equations and in formulas—that is, in the language of mathematical models.

However, as a practical matter, we do ultimately need numbers to analyze and design structures, for loads, for structural dimensions, for material properties, for particular behaviors, and for member sizes. As long as our problem is relatively small, we can use mathematical formulas to develop models of our structural problem and then calculate the numbers we need. For large-scale structural engineering problems, however, we need more efficient ways of generating meaningful numbers. These more efficient ways use computer programs that further refine our mathematical models by *discretizing* them—that is, by representing them in terms of physically meaningful quantities that can be associated with discrete points in a body or a structure. Several different ways of doing this sort of discretization of models were originally developed in terms of the continuous mathematics of elastic bodies, although they generally have their roots in the notion of casting a continuous function in the form of a *Fourier series* whose amplitudes represent values of the relevant dependent variable at specific, discrete locations. We will illustrate this discretization process at several points in what follows, especially as we seek approximate solutions to problems involving both bars and beams.

The most widely used, modern form of discretization, which goes well beyond what is possible with Fourier series, is the finite element method (FEM). Such approaches can be elegantly written in matrix form. Further, since matrix manipulation is easily automated, it too is a part of the FEM process.

Now, given the discursive nature of this book, we will deal with structures, loads, and geometry of limited complexity. There are extremely complicated issues that derive from attempts to achieve great accuracy for structural models that require thousands and or tens of thousands of unknowns, issues we do not address. Nor do we intend to touch many effects that are quite important in structural engineering, such as nonlinear materials and geometry, vibration and dynamics, structural stability, and so on. However, the approaches we will demonstrate can be taken as a model of good habits of thought for solving problems in structural engineering either analytically or computationally.

Bibliography

Dym, C. L. 1997. *Structural modeling and analysis.* New York: Cambridge University Press.

3

Exploring Intuition: Beams, Trusses, and Cylinders

Summary

Physical intuition is really important for successful modeling of structural behavior, and yet good intuition often develops with repeated application of structural models. In this chapter we explore three well-known structural models—elementary beams, trusses, and pressurized cylinders—with the aim of pointing out features to look for to test and develop intuition. First, we will discuss the bending of beams, looking at a variety of issues concerning the distribution of forces, stresses, and displacements. We will also examine a simple truss as an archetypal beam. Then we conclude by examining the displacements in thick cylinders under internal and external pressures to show results that may not be readily anticipated. These analyses provide an opportunity to explore and perhaps enhance our physical intuition.

Introduction

Courses in the strength of materials always develop the elementary theory of beams and almost always include studies of the effects of internal and external pressure on thin, shell-like circular cylinders. In advanced courses in the strength of materials and/or applied elasticity, analyses of both of these structures are readily developed from two-dimensional elasticity theory, with the intent of exposing the details of the stress and displacement distributions through the thicknesses of the beams and cylinders. Thus, we will start with some planar elasticity in Cartesian coordinates to discuss Euler–Bernoulli beam theory, and later we shall develop similar results in polar coordinates to do the same for thick-walled cylinders. The latter results will also be useful to us in Chapter 4 when we develop some models of arch behavior.

Engineering Beams: The Two-Dimensional Model

We begin by recalling that the basic purpose of elementary beam theory, tra-
ditionally known as Euler–Bernoulli beam theory to commemorate its cre-
ators, is to model the behavior of long slender elements that are loaded in a
direction orthogonal to the beam's long axis. Thus, a beam is a "planar" object
of length L and height h in the (x, z) plane, and the only loads considered are
represented as a downward transverse load per unit length of the top surface,
$q(x)$ (Figure 3.1). The beam's third dimension, normal to the plane of the paper,
is denoted by y, and the beam is assumed to have a thickness b and to be
unloaded in that direction. Assuming that there are no body forces (i.e., forces
per unit volume), the two-dimensional equations of equilibrium are

$$\frac{\partial \sigma_{xx}}{\partial x} + \frac{\partial \tau_{zx}}{\partial z} = 0$$

$$\frac{\partial \tau_{xz}}{\partial x} + \frac{\partial \sigma_{zz}}{\partial z} = 0$$

(3.1)

Before solving Equation (3.1), we recall that we have stipulated that our
beams are expected to have a thickness h that is much smaller than its length
L. To assess the consequences of this expectation, we recast Equation (3.1) into
a semidimensionless form by defining two dimensionless coordinates, $\xi \triangleq x/L$
and $\eta \triangleq z/h$:

$$\frac{\partial \sigma_{xx}}{\partial \xi} + \left(\frac{L}{h}\right)\frac{\partial \tau_{zx}}{\partial \eta} = 0$$

$$\frac{\partial \tau_{xz}}{\partial \xi} + \left(\frac{L}{h}\right)\frac{\partial \sigma_{zz}}{\partial \eta} = 0$$

(3.2)

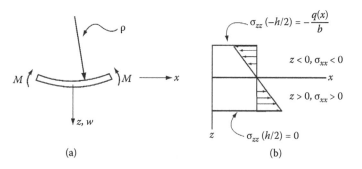

(a)   (b)

FIGURE 3.1
Elementary beam nomenclature and sign convention: (a) for the entire beam, showing the
moment and the radius of curvature; (b) the distribution of the bending stress through the
thickness.

Equation (3.2) immediately tells us something about the magnitudes of the stresses—that is,

$$\frac{|\tau_{zx}|}{|\sigma_{xx}|} = O\left(\frac{h}{L}\right)$$

$$\frac{|\sigma_{zz}|}{|\tau_{xz}|} = O\left(\frac{h}{L}\right)$$

(3.3)

This simple analysis of dimensions leads us to expect that the ratio of the magnitudes of the transverse shear stress to the axial normal stress will be about h/L, as would the ratio of the magnitudes of the normal stress through the thickness to the shear stress. Taken in combination, Equation (3.3) also says that the ratio of the magnitudes of the two normal stresses is $|\sigma_{zz}|/|\sigma_{xx}| = O(h/L)^2$. Thus, when anticipating results for *thin* beams for which we postulate $h/L \ll 1$, we should expect (or intuit) that $\sigma_{xx} \gg \tau_{xz} \gg \sigma_{zz}$. Remember, however, that Equation (3.3) does not assume anything about how big h and L actually are, so it is quite possible to use Equation (3.3) or any ensuing results to consider short, fatter structures!

There are several ways to solve the equations of equilibrium (Equation 3.1). For our purposes, we build on familiar results beam theory, although we take a nonstandard approach so as to illuminate some relative size arguments. We start by noting that the axial normal and shear stresses can be represented in terms of moment and shear force resultants, $M(x)$ and $V(x)$, respectively, and we restrict the details to beams having rectangular cross sections:

$$\sigma_{xx}(x,z) = \frac{M(x)z}{I}$$

$$\tau_{xz}(x,z) = \frac{V(x)}{2I}\left(\left(\frac{h}{2}\right)^2 - z^2\right)$$

(3.4)

where the stress resultants are defined as integrals over the beam's cross section,

$$M(x) = \int_{-h/2}^{h/2} \sigma_{xx}(x,z)zb\,dz$$

$$V(x) = \int_{-h/2}^{h/2} \tau_{xz}(x,z)b\,dz$$

(3.5)

and I is the second moment of the cross-section area (unfortunately and wrongly often called the moment of inertia),

$$I = \int_{-h/2}^{h/2} z^2 b\, dz \tag{3.6}$$

Thus, here, the beam cross section is rectangular with $A = bh$ and $I = bh^3/12$.

We substitute the stresses (Equation 3.4) into the first of the equilibrium equations (Equation 3.1) to find

$$\frac{dM(x)}{dx}\left(\frac{z}{I}\right) - \frac{V(x)z}{I} = 0 \tag{3.7}$$

which exposes a familiar and dimensionally correct relationship that satisfies equilibrium along the axis of the beam:

$$V(x) = \frac{dM(x)}{dx} \tag{3.8}$$

If we substitute those stresses into the second equation of equilibrium, we find here that

$$\frac{1}{2I}\frac{dV(x)}{dx}\left(\left(\frac{h}{2}\right)^2 - z^2\right) + \frac{\partial \sigma_{zz}(x,z)}{\partial z} = 0$$

which, when integrated with respect to z, yields

$$\sigma_{zz}(x,z) = \frac{V'(x)}{2I}\left(\frac{z^3}{3} - \frac{h^2 z}{4}\right) + f(x) \tag{3.9}$$

Note here that Equation (3.9) contains both an unknown function of integration, $f(x)$, as well as the shear force—whose determination is not yet clear. However, both of these unknowns will become known when we satisfy stress boundary conditions on the beam's loaded (upper) and unloaded (lower) surfaces—namely:

$$\sigma_{zz}\left(x, -\frac{h}{2}\right) = -\frac{q(x)}{b} \quad \text{and} \quad \sigma_{zz}\left(x, \frac{h}{2}\right) = 0 \tag{3.10}$$

Then the normal stress through the beam's thickness follows as

$$\sigma_{zz}(x,z) = \frac{V'(x)}{4b}\left(2 - 3\left(\frac{z}{h/2}\right) + \left(\frac{z}{h/2}\right)^3\right)$$ (3.11)

and the shear force resultant is determined from equilibrium in the transverse direction by

$$\frac{dV(x)}{dx} + q(x) = 0$$ (3.12)

Then the familiar transverse equilibrium equation of classical beam theory emerges from the combination of Equations (3.8) and (3.12):

$$\frac{d^2 M(x)}{dx^2} + q(x) = 0$$ (3.13)

Finally, with the aid of Equation (3.8), all three stress components can be expressed as functions of the moment $M(x)$ in both dimensional and dimensionless coordinates:

$$\sigma_{xx}(x,z) = \frac{M(x)z}{I} \equiv \frac{M(\xi)\eta}{Ah/12}$$

$$\tau_{xz}(x,z) = \frac{M'(x)}{2I}\left(\frac{h^2}{4} - z^2\right) \equiv \frac{3M'(\xi)}{2AL}\left(1 - 4\eta^2\right)$$ (3.14)

$$\sigma_{zz}(x,z) = \frac{d^2 M(x)/dx^2}{4b}\left(2 - 3\left(\frac{z}{h/2}\right) + \left(\frac{z}{h/2}\right)^3\right) \equiv \frac{hM''(\xi)}{4AL^2}\left(2 - 6\eta + 8\eta^3\right)$$

Equation (3.14) makes it quite clear that the anticipated magnitudes of the various stress ratios are exactly as anticipated; that is,

$$\frac{|\tau_{zx}|_{max}}{|\sigma_{xx}|_{max}} \propto \frac{hM'(\xi)}{4LM(\xi)} = O\left(\frac{h}{L}\right)$$

$$\frac{|\sigma_{zz}|_{max}}{|\sigma_{xx}|_{max}} \propto \frac{h^2 M''(\xi)}{6L^2 M(\xi)} = O\left(\frac{h}{L}\right)^2$$ (3.15)

Now, while the preceding results give us some things to think about when we analyze or design beams (or other structures, for which similar reasoning can be developed), they are focused on the stresses. In fact, we can also develop similar characterizations for the beam's deflections, as follows. Recall first that we model the deflections of real physical beams by expressing them in terms of the downward, transverse displacement of the beam's *neutral axis*, which is typically drawn longitudinally through the centroids of the beam's cross section. For the rectangular cross sections we are modeling, that axis is at the vertical center, and the positive downward deflection is denoted as $w(x)$. However, since the beam bends as it deflects, points above and below the neutral axis move axially according to the fundamental expression of the *Euler–Bernoulli hypothesis:*

$$u(x,z) \triangleq -z\frac{dw(x)}{dx} \tag{3.16}$$

Equation (3.16) also reflects the other oft-stated version of this hypothesis—namely, that *plane sections remain plane*. We can see this (namely, Figure 3.2) because a line element originally vertical (or normal to the beam axis) will only rotate through the *Euler angle* $(-dw(x)/dx)$ as the beam deflects an amount $w(x)$. It is also reflected in the three components of strain that result from the two components of displacement:

$$\varepsilon_{xx} = \frac{\partial u(x,z)}{\partial x} = -z\frac{d^2w(x)}{dx^2}, \quad \varepsilon_{zz} = \frac{\partial w(x)}{\partial z} = 0, \quad \gamma_{zx} = \frac{\partial u(x,z)}{\partial z} + \frac{\partial w(x)}{\partial x} \equiv 0 \tag{3.17}$$

It is also interesting to observe what Equation (3.16) and the last part of Equation (3.17) say about the beam's physical behavior at a clamped or fixed support. We take for granted that a clamped condition $dw(x)/dx = 0$ at a support means that the beam's axis remains horizontal. But by virtue of Equation (3.16) it also means that a line normal to the beam's axis undergoes no displacement in the axial direction, so it remains vertical there—another manifestation of plane sections remaining plane.

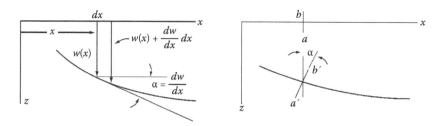

FIGURE 3.2
The displacement assumptions underlying the Euler–Bernoulli hypothesis of elementary beam theory.

With the strains known by Equation (3.17), we need only the relevant stress–strain law to calculate stresses, after which we can relate stresses and stress resultants to the beam's deflection and thus calculate its deflected shape. Since we assume the beam to be unloaded in the y direction and the depth b to be small, the appropriate version of Hooke's law here is that, for plane stress in the (x, z) plane,

$$\varepsilon_{xx} = \frac{1}{E}(\sigma_{xx} - v\sigma_{zz}), \quad \varepsilon_{zz} = \frac{1}{E}(\sigma_{zz} - v\sigma_{xx}), \quad \gamma_{xz} = \frac{\tau_{xz}}{G} \qquad (3.18)$$

Equation (3.18) presents some interesting modeling issues, as it is not immediately clear how to reconcile the strains developed in Equation (3.17) with the stresses written in Equation (3.14). One issue is whether we need (or not) to retain the transverse normal stress $\sigma_{zz}(x, z)$ here, inasmuch as we have already seen that it will be two orders of magnitude smaller than its axial counterpart. The first of Equation (3.18) certainly suggests that it will not much affect the value of the axial normal strain $\varepsilon_{xx}(x, z)$, especially since the Poisson ratio is such that $0 \leq v \leq 0.5$. Thus, we would then take $\sigma_{xx} = E\varepsilon_{xx}$ as our simple, one-dimensional stress–strain law.

A second issue of note is that the transverse normal strain $\varepsilon_{zz}(x, z)$, while nontrivial, plays no role in our equilibrium or other calculations. In any event, it can simply be regarded as a modest amount of transverse Poisson expansion that is of little account.

The third and last issue is more intriguing. The last part of Equation (3.17) says that the transverse shear strain is zero (a direct consequence of the Euler–Bernoulli hypothesis); a transverse shear stress is needed to maintain equilibrium, while the last of Equation (3.18) says that zero strain is inconsistent with a nonzero shear stress. In fact, we accept this inconsistency because our beam models are very successful, both in experiments and in practice. We describe the inconsistency by saying that a beam is *infinitely rigid or stiff in shear,* which then allows us to accept a finite shear stress that produces a vanishing shear strain. We will also show in Equation (3.32), when we discuss the relative sizes of the beam's displacement field, that the shear strain is in fact a very, very small quantity.

We now use the relationship (Equation 3.17) between strain and displacement together with our simple one-dimensional stress–strain law to express the moment and shear resultants (from the definitions: Equation 3.5) in terms of the beam deflection:

$$M(x) = -EI\frac{d^2w(x)}{dx^2}$$

$$V(x) = -EI\frac{d^3w(x)}{dx^3} \qquad (3.19)$$

And with the aid of the first part of Equation (3.19), often called the moment-curvature relation, to cast the transverse equilibrium Equation (3.13) in terms of the beam deflection,

$$EI\frac{d^4w(x)}{dx^4} = q(x) \tag{3.20}$$

Now we have two versions of the equation of equilibrium for a beam bent under transverse loading: the first (Equation 3.13) is expressed in terms of the moment $M(x)$ and can be used for any statically determinate beam, while the second (Equation 3.20) is expressed in terms of the displacement $w(x)$ and can be used for any beam, determinate or indeterminate. All that remains to formulate a complete problem statement are the appropriate boundary conditions.

The most efficient and general way to generate the boundary conditions attendant to a specific structural model is to apply the "principle of minimum (total) potential energy" by considering the variations of that total potential energy as the displacements are varied. For the beam, this means considering the effect of displacement variations on the total potential energy

$$\Pi = \frac{EI}{2}\int\limits_0^L \left(\frac{d^2w(x)}{dx^2}\right)^2 dx - \int\limits_0^L q(x)\delta w(x)dx \tag{3.21}$$

When such variations $\delta w(x)$ are implemented, the first variation of the total potential energy is found to be

$$\delta^{(1)}\Pi = \int\limits_0^L EIw''(x)\delta w''(x)dx - \int\limits_0^L q(x)\delta w(x)dx \tag{3.22}$$

Before setting this first variation of the total potential energy to zero, as required by the minimum energy principle, we integrate by parts twice, assuming as always that the delta (variational) and derivative operators commute. After the integrations, we obtain the first variation of the total potential as

$$\delta^{(1)}\Pi = \left[EIw''(x)\delta w'(x)\right]_0^L - \left[\frac{d(EIw''(x))}{dx}\delta w(x)\right]_0^L + \int\limits_0^L\left[\frac{d^2(EIw''(x))}{dx^2} - q(x)\right]\delta w(x)dx \tag{3.23}$$

Recognizing Equation (3.19), which defines the moment and the shear resultants in terms of the bending deflection, we can rewrite Equation (3.23) as

$$\delta^{(1)}\Pi = -\left[M(x)\delta w'(x)\right]_0^L + \left[V(x)\delta w(x)\right]_0^L - \int_0^L \left[M''(x) + q(x)\right]\delta w(x)\,dx \quad (3.24)$$

The equation of equilibrium for beam bending results from the vanishing of the integral as we set the first variation (Equation 3.24) to zero is

$$\frac{d^2 M(x)}{dx^2} + q(x) = 0 \quad (3.25)$$

which is exactly the same as our previously derived Equation (3.13). The corresponding boundary conditions are found by setting the bracketed terms in the equation to zero at the beam's two ends, $x = 0$ and $x = L$. This provides a total of four boundary conditions needed for the fourth-order differential Equation (3.20):

$$\text{Either} \quad M(x) = 0 \quad \text{or} \quad \delta w'(x) = 0 \quad (3.26)$$

and

$$\text{Either} \quad V(x) = 0 \quad \text{or} \quad \delta w(x) = 0 \quad (3.27)$$

Equations (3.26) and (3.27) represent a pair of *generalized force–displacement dualities;* that is, at a given boundary point we prescribe (1) either a moment or a corresponding slope *and* (2) either a shear force or a corresponding displacement. In other words, at each point we have a pair of choices of choosing between *kinematic* or *geometric* (e.g., slope, displacement) and *force* (e.g., moment, shear) conditions.

Having now derived the equilibrium equation in terms of the displacement and the boundary conditions needed to solve it, we also note that this derivation did not provide the same opportunities to reason about the order of magnitude of the displacement itself. That is due in part to the fact that the most basic assumption about the relative size of displacements in typical elastic structures actually stems from the assumption that linear elastic models can be used to describe them.

A linear elastic model has two components: a linear stress–strain law (namely, Equation 3.18), which we have certainly seen previously, and a linear relation between strain and displacement, which we also saw in Equation (3.17). But the latter assumption rests on a tight restriction that derives from the fact that a completely general kinematic description of strain in an elastic body requires nonlinear terms (which form the basis for nonlinear elasticity theory, including stability theory). It can be shown that the nonlinear terms in the strain-displacement relations can only be ignored when, in this instance, the beam's bending deflection is small compared with its thickness—that is,

$$\frac{|w(x)|_{max}}{h} < 1 \qquad (3.28)$$

This assumption was implicitly incorporated when we wrote down the Euler–Bernoulli hypothesis (Equation 3.17).

We can make use of Equation (3.28) to develop a guideline for the magnitude of the stresses (beyond the relative size ordering given in Equation 3.15). By combining the first parts of Equations (3.14) and (3.19), respectively, we can calculate the bending (axial) stress in terms of the curvature as

$$\sigma_{xx}(x,z) = \left(-EI\frac{d^2w(x)}{dx^2} \right)\left(\frac{z}{I} \right) = -Ez\frac{d^2w(x)}{dx^2} \qquad (3.29)$$

In the light of Equation (3.28), by rewriting Equation (3.29) in terms of dimensionless coordinates we can also see that

$$\frac{|\sigma_{xx}(x,z)|_{max}}{E} \propto \left(\frac{z}{h} \right)\left(\frac{|w(x)|_{max}}{h} \right)\left(\frac{h}{L} \right)^2 = O\left(\frac{h}{L} \right)^2 \qquad (3.30)$$

We should thus expect that the bending stress we calculate should always be at least two orders of magnitude smaller than the modulus of elasticity of the beam's material. Of course, this does not mean ignoring yield or failure stresses, but is simply an intuitive guideline against which to eyeball or benchmark numerical calculations.

A similar calculation can be made for the shear stress and its corresponding "zero" shear strain. Here we combine the second parts of Equations (3.14) and (3.19), respectively, to calculate the shear strain and stress in terms of derivatives of the curvature

$$\gamma_{xz}(x,z) = \frac{\tau_{xz}(x,z)}{G} = \left(\frac{EI}{2GI}\frac{d^3w(x)}{dx^3} \right)\left(z^2 - \frac{h^2}{4} \right) \qquad (3.31)$$

Again, in the light of Equation (3.28), we now write Equation (3.31) as a comparison (in terms of dimensionless coordinates) of the magnitude of the shear strain to the magnitude of the Euler angle:

$$\frac{|\gamma_{xx}(x,z)|_{max}}{|w'(x)|_{max}} \propto \left(\frac{1+v}{4} \right)\left(\frac{h}{L} \right)^2 = O\left(\frac{h}{L} \right)^2 \qquad (3.32)$$

FIGURE 3.3
A classical cantilever beam with a load *P* at its tip.

Thus, by accepting a zero shear strain, we are admitting a miniscule error that is two orders of magnitude smaller than the Euler angle, which is itself extremely small.

One more note on relative size in and intuitive reasoning about beams. A full, formal solution for the displacement of the neutral axis of a cantilever beam subjected to a tip load *P* (Figure 3.3) can be found to be as follows, subject to the clamping condition that the vertical plane at the clamp remains planar (recall our earlier remark on this point with regard to the Euler–Bernoulli hypothesis):

$$w(x,z=0) = \frac{PL^3}{6EI}\left[3\left(\frac{x}{L}\right)^2 - \left(\frac{x}{L}\right)^3 + \frac{3(1+\nu)}{2}\left(\frac{h}{L}\right)^2\left(\frac{x}{L}\right)\right] \tag{3.33}$$

The first bracketed term in Equation (3.33) should be recognized as the classical elementary solution for the deflected shape of a tip-loaded cantilever. Thus, for long slender beams, it is very easy to see that the second term in the brackets may be safely discarded or ignored. That second term can be rewritten to make its shear component more explicit by noting the definitions of the shear modulus *G* and of the area *A* and second moment of the area *I*:

$$\frac{PL^3}{6EI}\left[\frac{3(1+\nu)}{2}\left(\frac{h}{L}\right)^2\left(\frac{x}{L}\right)\right] = \frac{3PL^3}{24GI}\left(\frac{h}{L}\right)^2\left(\frac{x}{L}\right) = \frac{3PL}{2GA}\left(\frac{x}{L}\right) \tag{3.34}$$

Then the cantilever's deflected shape becomes

$$w(x,z=0) = \frac{PL^3}{6EI}\left(3\left(\frac{x}{L}\right)^2 - \left(\frac{x}{L}\right)^3 \right) + \frac{3PL}{2GA}\left(\frac{x}{L}\right) \tag{3.35}$$

In this form of the cantilever's deflection, it is not at all obvious that the bending term is significantly smaller than the shear contribution. Of course, we can recall the argument that a thin beam is infinitely rigid in shear, so allowing *G* → ∞ would dictate discarding the second term. But it is easier to make the argument when all the terms have as much of a common factor as possible, as in Equation (3.33).

This also raises another point. We have focused all along on modeling long slender beams: We have particularly called out the thickness h and its size relative to the length L as important quantities. We have also cast our results in terms of EI. Yet we have only taken the ratio h/L to be small in citing a few specific results, and we have just seen that the apparent common factor can be revised. So, taking the analysis of Equation (3.34) one step further, we can rewrite Equation (3.33) with a new common factor, namely,

$$w(x, z = 0) = \frac{PL}{GA}\left[\frac{1}{(1+v)}\left(\frac{L}{h}\right)^2\left(3\left(\frac{x}{L}\right)^2 - \left(\frac{x}{L}\right)^3\right) + \frac{3}{2}\left(\frac{x}{L}\right)\right] \quad (3.36)$$

In this third form, it is still possible to see that the first term dominates for long slender beams, yet its relationship to the bending problem is not exactly obvious. On the other hand, Equation (3.36) does suggest that for very short stubby beams, for which $L/h < 1$, the second term will be more important, and we can clearly identify that dominant term as due to shear.

Reasoning Intuitively about Engineering Beams

We noted that we can enhance our intuition by using the kinds of dimensional thinking we have outlined previously. We want to build further on the importance of intuition, now looking at some issues that relate to approximation and to our visual intuition. And, not coincidentally, while we are working in traditional closed-form mathematics, these issues are both central to the successful computer modeling of structures.

In the category of visual thinking, consider the beam shown in Figure 3.4. Note that it has a central concentrated load P and an additional support at $x = l$ that makes the beam statically indeterminate. The intuition question is, "How does this beam behave when the middle support moves to

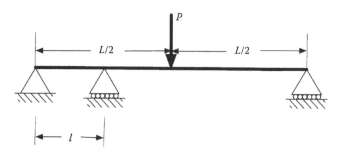

FIGURE 3.4
An indeterminate beam subject to a concentrated midpoint load P, with a middle support located at a distance l from the left-hand pin support.

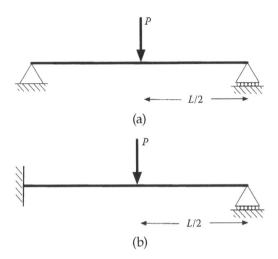

FIGURE 3.5
The two possible limiting cases of the interminate beam of Figure 3.4: (a) a centrally loaded simple (determine) beam; and (b) a centrally loaded interminate beam that is fixed at the left and simply suported at the right end.

the left—that is, when $l \to 0$?" Two alternative possibilities are shown in Figure 3.5; which is the right limit case? In the first instance, it would appear that the movable pin support simply overlaps and duplicates the pin support at the beam's left end. In the second possibility, we might argue that the beam's deflection in that interval becomes very small and that it flattens out as the middle moving pin closes in on the end pin. This suggests a clamp at the beam's left end. So, which is it?

We can answer this question by considering the statically determinate beam of Figure 3.5(b). The deflections under the loads R and P can be expressed in terms of standard *influence coefficients*, f_{ij}, as

$$\delta_R \triangleq w(l) = f_{11}R + f_{12}P$$

$$\delta_P \triangleq w(L/2) = f_{21}R + f_{22}P$$

(3.37)

where, in this instance, the influence coefficients are

$$f_{11} = \frac{L^3}{3EI}\left(\frac{l}{L}\right)^2\left(1-\frac{l}{L}\right)^2$$

$$f_{12} = \frac{L^3}{16EI}\left(\frac{l}{L}\right)\left[1-\frac{4}{3}\left(\frac{l}{L}\right)^2\right] = f_{21}$$

(3.38)

$$f_{22} = \frac{L^3}{48EI}$$

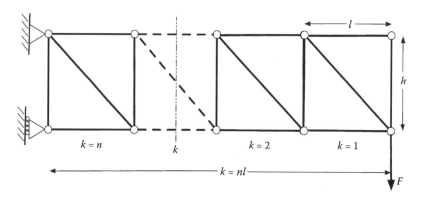

FIGURE 3.6
Alternative limit cases when the middle support of the beam shown in Figure 3.4 moves to the left support as $l \to 0$: (a) a simply supported beam; (b) a beam clamped at one end and pinned at the other.

Note that the symmetric off-diagonal influence coefficients f_{12} and f_{21} clearly illustrate the well-known Maxwell reciprocal theorem. Now, placing a support at $x = l$ requires that

$$\delta_P = 0 \quad \Rightarrow \quad R = -\left(\frac{f_{12}}{f_{11}}\right) P \tag{3.39}$$

which thus means that

$$\delta_P = \left(\frac{f_{11} f_{22} - f_{12}^2}{f_{11}}\right) P \tag{3.40}$$

If the influence coefficients (Equation 3.38) are substituted into Equation (3.40), that deflection becomes

$$\delta_P = \frac{PL^3}{48EI}\left[1 - \frac{9}{16}\frac{\left(1 - \frac{4}{3}\left(\frac{l}{L}\right)^2\right)}{\left(1 - \frac{l}{L}\right)^2}\right] \tag{3.41}$$

If we now take the limit of the deflection (Equation 3.41) as $l/L \to 0$, we find that

$$\delta_P\big|_{l \to 0} = \frac{7PL^3}{768EI} \tag{3.42}$$

Equation (3.42) is the central deflection corresponding to the fixed-pinned beam of Figure 3.5(b). Thus, this result suggests that the model of Figure 3.5(b) is the correct guide for our intuition as it focuses on how the deflection magnitude would change as we move the middle support ever closer to the left-hand pin.

This is not to say that the elegance of the preceding analysis does not raise further questions. In particular, if we substitute the influence coefficients (Equation 3.38) into the conclusion (Equation 3.39), we find the following support reaction at $x = l$:

$$R = -\frac{3}{16} \frac{\left[\left(1 - \frac{4}{3}\left(\frac{l}{L}\right)^2\right)\right]}{\left[\left(\frac{l}{L}\right)\left(1 - \frac{l}{L}\right)^2\right]} P \tag{3.43}$$

Equation (3.43) seems to signify a problem with our intuition as the reaction R is singular in that it grows infinitely large as $l/L \to 0$. However, we should keep two points in mind. One is that the shear resultants under concentrated loads on beams exhibit singular behavior as they pass through those loaded points. The second point is that we should recognize a guide for our intuition: Namely, imagine the moment needed to bend a very short beam. It is easy to see that the couple formed at the left end is proportional to the product $R \cdot l$, in which case Equation (3.43) shows that product to be finite. We could confirm this intuition by trying to bend ever-shortened model beams.

As a second category of approximation, consider the simple beam shown in Figure 3.7. That simple beam is loaded with three equally spaced (at the beam's quarter points) concentrated loads $P/3$. For beams and structures generally, we make a major assumption when we identify loads as concentrated: It is simply not physically possible to apply a force at a point because the very notion of a point is a mathematical idealization, as is the idea that a force can be applied over an infinitesimally small area. A second level

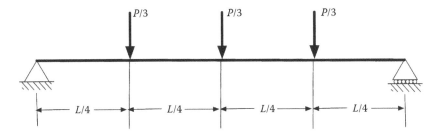

FIGURE 3.7
A simple beam with three equally spaced loads $P/3$.

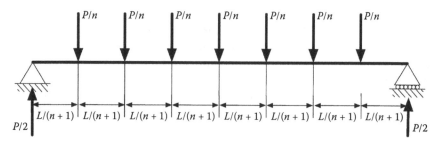

FIGURE 3.8
A simple beam with n loads P/n at uniform spacing $a_n = L/(n+1)$.

of idealization is that we can ignore the very fine details of the stresses in the beam at points under or very near a concentrated load. In fact, it can be shown that the stresses described by Euler–Bernoulli beam theory are extremely accurate except in the immediate neighborhood of that concentrated load.

For our purposes, though, an interesting question is, "How should we handle a case where there are several concentrated loads that are spaced rather closely together?" Figure 3.8 shows an obvious generalization of the situation of Figure 3.7: There are n loads P/n, at a spacing $a_n = L/(n+1)$, with a total load of P and equal end reactions $P/2$. Note that as n gets large, the spacing grows smaller as $1/n$, which suggests that we might aggregate the individual loads into a constant uniform load $q(x) = q_0 = (P/n)/(L/(n+1)) \cong P/L$. This is dimensionally correct for a uniform line load per unit length of beam, and it conforms with our intuitive expectations. But can we prove that intuition is correct? Further, can we find any easily applicable predictors when there are just a few loads, as in Figure 3.7?

We answer these two questions by first considering the simple beam of Figure 3.9 subject to a single concentrated load P at $x = a$. The deflection of any point x along the beam is easily found to be

$$w(x; P@a) = \frac{PL^3}{6EI}\left[\left(\frac{b}{L}\right)\left(1 - \left(\frac{b}{L}\right)^2\right)\left(\frac{x}{L}\right) - \left(\frac{b}{L}\right)\left(\frac{x}{L}\right)^3 + \left(\frac{x}{L} - \frac{a}{L}\right)^3 H(x-a)\right] \quad (3.44)$$

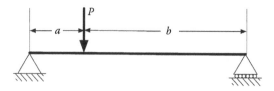

FIGURE 3.9
A determinate beam subject to a concentrated load P located at a distance a from the left-hand pin support. Note that $a + b = L$.

Note that we can use Equation (3.44) to calculate the influence coefficients of Equation (3.37) by superposing

$$\delta_R = w(a; R@a) + w(a; P@b)$$

$$\delta_P = w(b; R@a) + w(b; P@b)$$

(3.45)

To characterize the response of the beam of Figure 3.8, with n loads P/n separated by a common distance of $L/(n+1)$, we will assume that the number of loads is odd and will calculate the midpoint (i.e., $x = L/2$) deflection. (Since we are interested in the limit $n \to \infty$, it makes no difference whether we assume that n is odd, even, or one of all integers.) For the kth load $P_k = P/n$ located at a distance $a_k = ka_n = kL/(n+1)$ from the left pin, with $b_k = L - a_k$ and a little algebra, we can use Equation (3.44) to find the midpoint deflection as

$$w_k\left(\frac{L}{2}; P/n@a_k\right) = \frac{PL^3}{16EI}\left[\left(\frac{a_k}{L}\right) - \left(\frac{4}{3}\right)\left(\frac{a_k}{L}\right)^3\right]$$

(3.46)

We find the midpoint deflection of the beam due to the entire odd sequence of loads P_k by summing Equation (3.46) over the range of odd integers $0 \leq k \leq (n-1)/2$, doubling that result to account for symmetry, and then, finally, adding the deflection due to the load P/n at the very center of the beam for the simple centrally loaded beam ($n = 1$), which we excluded by choosing to represent the odd numbers as $k = (n-1)/2$. Thus,

$$w_T\big|_{L/2} = \frac{(P/n)L^3}{48EI} + 2\sum_{k=1}^{\frac{n-1}{2}} w_k\left(\frac{L}{2}; P/n@a_k\right)$$

$$= \frac{PL^3}{48EIn}\left[1 + \left(\frac{6}{n+1}\right)\sum_{k=1}^{\frac{n-1}{2}} k - \left(\frac{2}{n+1}\right)^3 \sum_{k=1}^{\frac{n-1}{2}} k^3\right]$$

(3.47)

Now, as daunting as Equation (3.47) may appear, the finite sums are known:

$$\sum_{k=1}^{\frac{n-1}{2}} k = \frac{n^2 - 1}{8} \quad \text{and} \quad \sum_{k=1}^{\frac{n-1}{2}} k^3 = \left(\frac{n^2 - 1}{8}\right)^2$$

(3.48)

With the sums (Equation 3.48), we find that the midpoint deflection due to the superposed set of loads of Figure 3.8 is

$$w_T\big|_{L/2} = \frac{PL^3}{384EI}\left[\frac{5n^2+10n+1}{n(n+1)}\right]$$
(3.49)

What can we learn from this elegant piece of analysis? First, we can now calculate the midpoint deflection due to an odd collection of loads P/n; that is,

$$w_T\big|_{L/2,n=1} = \frac{PL^3}{48EI}$$

$$w_T\big|_{L/2,n=3} = \frac{PL^3}{384EI}\left[\frac{45+30+1}{3(4)}\right]$$
(3.50)

$$w_T\big|_{L/2,n=5} = \frac{PL^3}{384EI}\left[\frac{125+50+1}{5(6)}\right]$$

Second, and perhaps most interesting, we can also take the limit $n \to \infty$ in Equation (3.49) to find out whether we do get the deflection due to a uniform load. That limit result is, simply,

$$w_T\big|_{L/2,n=1} = \frac{5PL^3}{348EI}$$
(3.51)

Equation (3.51) is exactly the midpoint deflection due to a uniform load $q_0 = P/L$, which confirms the well-known intuition that a large enough sequence of concentrated loads elicits the same response as a uniform load.

We can also use the preceding results to frame and answer another question, which may be useful for its practical applications: Can we similarly calculate a beam's stresses when the loads of Figure 3.8 are applied? Note that we cannot determine the stresses directly nor can we appropriately differentiate the results (Equation 3.49). However, we can calculate the moment $M(x)$ at the beam's center and use that to calculate the midpoint axial stress using the classic beam formula (the first of Equation 3.14). (We could similarly calculate the shear force $V(x)$ to determine the shear stress through the thickness, were we so inclined.) For the midpoint moment, from a simple free-body diagram (namely, Figure 3.10), we find

$$M_k\left(\frac{L}{2};P/n@a_k\right) = \frac{PL}{4} - \frac{Pa_k}{n} \equiv \frac{PL}{4} - \left(\frac{P}{n}\right)\left(\frac{kL}{n+1}\right)$$
(3.52)

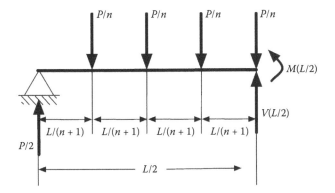

FIGURE 3.10
The free-body diagram of the left half of the simple beam with n equally spaced ($a_n = L/(n + 1)$) loads P/n depicted in Figure 3.8.

Again, assuming an odd number of loads and using the first of the sums (Equation 3.48), we can easily find that the midpoint moment due to a series of loads P/n is

$$M\left(\frac{L}{2}\right)\bigg|_n = \frac{PL}{4} - \left(\frac{PL}{8}\right)\left(\frac{n-1}{n}\right) \tag{3.53}$$

For smaller numbers of loads, Equation (3.53) shows that

$$M\left(\frac{L}{2}\right)\bigg|_{n=1} = \frac{PL}{4} - \left(\frac{PL}{8}\right)\left(\frac{1-1}{1}\right) = \frac{PL}{4}$$

$$M\left(\frac{L}{2}\right)\bigg|_{n=3} = \frac{PL}{4} - \left(\frac{PL}{8}\right)\left(\frac{2}{3}\right) = \frac{PL}{6} \tag{3.54}$$

$$M\left(\frac{L}{2}\right)\bigg|_{n=5} = \frac{PL}{4} - \left(\frac{PL}{8}\right)\left(\frac{4}{5}\right) = \frac{3PL}{20}$$

The first part of Equation (3.54) is the well-known result for a simple beam with a point load P at its midpoint, while the second and third results show a decreasing midpoint moment as the number of loads P/n is increased with n. In the limit $n \to \infty$, the midpoint moment becomes

$$M\left(\frac{L}{2}\right)\bigg|_{n\to\infty} = \frac{PL}{4} - \frac{PL}{8} = \frac{PL}{8} \tag{3.55}$$

Equation (3.55) is exactly the midpoint moment of a simple beam subjected to a uniform line load $q_0 = P/L$, thus confirming once again our intuition about the transition from discrete to uniform loading.

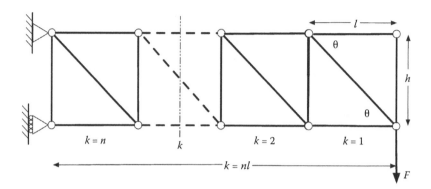

FIGURE 3.11
A simple cantilevered truss that is "fixed" at its root and supports a concentrated load F at its free end.

Trusses as Beams

We now turn to trusses in order to display another form of beam behavior. Simply put, we will examine a cantilevered truss under a tip load F and show how its behavior is remarkably like the beam models we described earlier. In particular, we will see how long trusses behave like classical Euler–Bernoulli beams and that, in shorter trusses, shear displacements become dominant.

We begin with the truss displayed in Figure 3.11. Note that this is about as simple a truss as we can develop by adding two bars and a connecting point to a starting triangle, which is how trusses are "assembled." The bars are assumed to support only axial loads; that is, they do not individually bend or support shear. It is only when these bars are organized into trusses that the resulting truss structure behaves like a beam. The truss is made up of n panels, each of length l and height h. The trusses' total length is thus $L = nl$. We define and show in Figure 3.11 an angle $\theta = \cot^{-1} l/h$, and we assume for simplicity that all of the truss bars are made of the same material and have a common area A. This is a simple, statically determinate external truss that is supported by one vertical and two horizontal reactions, as depicted in Figure 3.11. It is also statically determinate internally, meaning that we can calculate all of the bar forces directly from the equations of static equilibrium.

In fact, this truss is so simple that, by taking a vertical section through any of the n panels (see Figure 3.12a)—say, the kth panel—we can immediately see from vertical equilibrium that all of the n diagonal bar forces F_{Dk} are all the same; that is,

$$F_{Dk} = F/\sin\theta, \quad k = 1, 2, \dots n \tag{3.56}$$

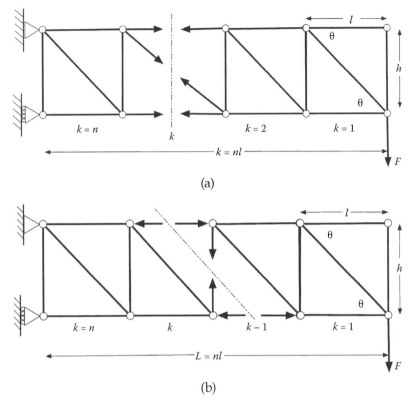

FIGURE 3.12
Sections taken through the elementary truss of Figure 3.10: (a) a vertical section through any of the n panels; (b) a diagonal section across any pair of adjoining panels. In each case, a net transverse force of F must be supported to satisfy transverse (vertical) equilibrium.

A section through any adjoining pair of panels (namely, Figure 3.12b) shows that each of the $n - 1$ internal vertical bars carries the same compressive load F_{Vk}:

$$F_{Vk} = -F, \quad k = 1, 2, \dots (n-1) \tag{3.57}$$

If we now sum forces at each joint along the truss's top chord, starting at the top right joint, it is easily seen that the top chord bar forces F_{TCk} are

$$F_{TCk} = kF \cot \theta, \quad k = 0, 1, \dots (n-1) \tag{3.58}$$

Note that each of the top chord bars is in tension (which is why they are positive!). A similar static analysis of the joints along the truss's bottom chord shows that those bottom chord bar forces F_{BCk} are in compression:

$$F_{BCk} = -kF \cot \theta, \quad k = 1, 2, \dots n \tag{3.59}$$

We also recognize that the bar lengths vary according to where they appear—that is,

$$L_{Dk} = l/\cos\theta, \quad L_{Vk} = h = l\tan\theta, \quad L_{TCk} = L_{BCk} = l = L/n \qquad (3.60)$$

To illustrate the beam behavior of the truss, we now calculate the deflection of the truss's loaded tip by applying Castigliano's second theorem. Thus, we write the total complementary energy of the truss as the sum of the complementary energies of the diagonal bars, the vertical bars, the bars in the top chord, and those in the bottom cord:

$$U^* = \sum_{k=1}^{n} \frac{F_{Dk}^2 L_{Dk}}{2EA} + \sum_{k=1}^{n-1} \frac{F_{Vk}^2 L_{Vk}}{2EA} + \sum_{k=0}^{n-1} \frac{F_{TCk}^2 L_{TCk}}{2EA} + \sum_{k=1}^{n} \frac{F_{BCk}^2 L_{BCk}}{2EA} \qquad (3.61)$$

If we recognize the bar forces given by Equations (3.56)–(3.59) and the corresponding lengths of each set of bar forces (Equation 3.60), we can write the complementary energy (Equation 3.61) as follows, noting that the first two sums are easily summed:

$$U^* = \frac{F^2 L}{2EA\sin^2\theta\cos\theta} + \frac{F^2(L-l)\tan\theta}{2EA} + \frac{F^2 L\cot^2\theta}{2EAn}\sum_{k=1}^{n}(k-1)^2 + \frac{F^2 L\cot^2\theta}{2EAn}\sum_{k=1}^{n}k^2$$

$$(3.62)$$

Interestingly enough, the last two finite series can be summed and those sums are

$$\sum_{k=1}^{n}(k-1)^2 = \frac{n(n-1)(2n-1)}{6}, \quad \sum_{k=1}^{n}k^2 = \frac{n(n+1)(2n+1)}{6} \qquad (3.63)$$

With the aid of the sums (Equation 3.63), we can then write the complementary energy simply as

$$U^* = \frac{F^2 L}{2EA}\left[\frac{1}{\sin^2\theta\cos\theta} + \left(1 - \frac{l}{L}\right)\tan\theta + \frac{(2n^2+1)\cot^2\theta}{3}\right] \qquad (3.64)$$

where the terms within the brackets of Equation (3.64) respectively reflect the diagonal, vertical, and top and bottom chord bars of the truss. If we eliminate the trigonometric functions in Equation (3.64) by recognizing their expressions in terms of the lengths l, L, and h, we can write the complementary energy in a physically more meaningful form:

$$U^* = \frac{F^2 L^3}{3EAh^2}\left[\frac{2n^2+1}{2n^2} + \frac{3(n-1)}{2n^3}\left(\frac{h}{l}\right)^3 + \frac{3}{2n^2}\left(1+\left(\frac{h}{l}\right)^2\right)^{3/2}\right] \qquad (3.65)$$

There is one more point before we apply Castigliano's second theorem to calculate the tip deflection from the complementary energy (Equation 3.65). If we were to look at the truss along its long axis, we would see top and bottom chord bar areas A at a distance $h/2$ from the vertical center of the truss, which means that we could approximate a second moment of the area of the truss cross section as $I = Ah^2/2$. Of course, this is a somewhat conservative estimate in that it does, to some extent, ignore the contributions of the diagonal and vertical bars to the truss's second moment. In effect, the diagonal and vertical bars serve as weightless, "area-less" spacers that serve only to bind the top and bottom chords to each other. This being assumed, the complementary energy (Equation 3.65) can be written as

$$U^* = \frac{F^2 L^3}{6EI}\left[\frac{2n^2+1}{2n^2} + \frac{3(n-1)}{2n^3}\left(\frac{h}{l}\right)^3 + \frac{3}{2n^2}\left(1+\left(\frac{h}{l}\right)^2\right)^{3/2}\right] \qquad (3.66)$$

and the tip deflection follows from Castigliano's second theorem as

$$\delta_{tip} = \frac{dU^*}{dF} = \frac{FL^3}{3EI}\left[\frac{2n^2+1}{2n^2} + \frac{3(n-1)}{2n^3}\left(\frac{h}{l}\right)^3 + \frac{3}{2n^2}\left(1+\left(\frac{h}{l}\right)^2\right)^{3/2}\right] \qquad (3.67)$$

The first term in the brackets stems from the extension of the top cord bars and the concomitant compression of the bottom chord bars. The term $(n-1)/n^3$ derives from the compression of the $n-1$ vertical members, while the $(1+(h/l)^2)^{3/2}$ term stems from the tension in the diagonal bars. Clearly, since the truss beam's length L is proportional to the number of panels n, a truss beam will grow longer and more slender as the number of panels increases. Equally clearly, the effect of the vertical and diagonal bars drops with the square of the number of panels. Indeed, for a relatively small number of panels—say, $n \geq 3$,

$$\delta_{tip}\big|_{n>3} = \frac{FL^3}{3EI}\left[1+O\left(\frac{1}{n^2}\right)\right] \cong \frac{FL^3}{3EI} \qquad (3.68)$$

That is, with even as few as three panels, the tip deflection of the truss beam is virtually identical, to within about 5%, to the corresponding

Euler–Bernoulli beam theory result! Conversely, with $n \leq 3$, the vertical and diagonal bars become much more important, with the diagonal bars (the last term in Equation 3.68), which support the transverse shear of the truss beam, becoming the most dominant structural element.

Pressurized Circular Cylinders: A Two-Dimensional Model

The effects of internal and external pressure on circular cylinders are well known. In the most elementary discussions we typically analyze very thin-shell structures, with our modeling aimed toward determining stress components that are uniform through the thickness. We can then show that the in-plane circumferential and axial stresses in cylinders are larger than the transverse normal stress by a factor proportional to the shell's radius-to-thickness ratio (R/h). In more advanced studies we focus on the stress distributions through the walls of thick cylinders, from which we can derive the limiting thin-shell results. The corresponding displacements (and strains) of thick-walled cylinders are not generally discussed because they are not usually a serious design issue.

However, it turns out that the consideration of displacements and the resulting geometry questions provide an interesting opportunity to test and enhance physical intuition. There are basic questions:

- For an internally pressurized thick cylinder, does a point on the loaded (inner) surface move more than or less than a point on the unloaded (outer) surface?

- For an externally pressurized thick cylinder, does a point on the loaded (outer) surface move more than or less than a point on the unloaded (inner) surface?

The answers to these two questions are not as obvious as it first appears, and they differ with the load (i.e., internal vs. external pressure) and vary with Poisson's ratio, v. For a shell made of an incompressible material (i.e., $v = 0.5$), the overall shell volume must be preserved, which for plane strain suggests that changes Δb of the outer radius b be accompanied by corresponding changes Δa of the inner radius a:

$$(\Delta a)_{cylinder} = \left(\frac{b}{a}\right)(\Delta b)_{cylinder} \tag{3.69}$$

Since $b/a \geq 1$, Equation (3.69) suggests that $\Delta b / \Delta a \leq 1$, independently of loading (i.e., regardless of which surface, inner or outer, is pressurized).

On the other hand, for $0 \le v \le 0.5$, the stress–strain relations are coupled and local volume changes can occur. Thus, the in-plane strains depend on both in-plane and transverse normal stress components, which are reflected in changes in the effective stiffnesses of circular cylindrical arches. Further, diffusion of the displacements from a loaded surface, a phenomenon we often refer to as *geometric spreading*, suggests that the *near* displacement of a loaded surface should be larger than the *far* displacement of the unloaded surface:

$$p_i \neq 0, \, p_o = 0: \quad \frac{\Delta a}{\Delta b} > 1$$

$$p_i = 0, \, p_o \neq 0: \quad \frac{\Delta b}{\Delta a} > 1$$

(3.70)

Clearly, we cannot answer the preceding questions very easily, and the answers are not intuitively obvious. So let us analyze a hollow circular cylinder of inner radius a and outer radius b, pressurized on its inner surface by a uniform pressure p_i and on its outer surface by a uniform pressure p_o (Figure 3.13). The equation of equilibrium (corresponding to Equation 3.1) is, in polar coordinates, for axisymmetric loading and absent any body forces:

$$\frac{d\sigma_{rr}}{dr} + \frac{\sigma_{rr} - \sigma_{\theta\theta}}{r} = 0 \tag{3.71}$$

Note that there is only one independent variable in Equation (3.71), so the kind of dimensional estimating we did for its Cartesian counterparts (Equation 3.2) is not going to be of much use here, other than to note that it is

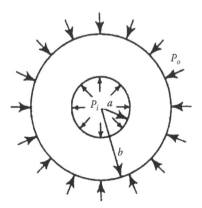

FIGURE 3.13
A thick cylinder loaded at its inner radius ($r = a$) with a pressure p_i and, at its outer radius ($r = b$), with a pressure p_o.

fairly clear from Equation (3.71) that the two stress components, σ_{rr} and $\sigma_{\theta\theta}$, are going to be of the same order of magnitude. In fact, we can easily look up and write down the radial and circumferential pressures for this problem, which readily confirm that they have similar magnitudes:

$$\sigma_{rr} = \frac{p_i a^2}{b^2 - a^2}\left(1 - \frac{b^2}{r^2}\right) - \frac{p_o b^2}{b^2 - a^2}\left(1 - \frac{a^2}{r^2}\right)$$

$$\sigma_{\theta\theta} = \frac{p_i a^2}{b^2 - a^2}\left(1 + \frac{b^2}{r^2}\right) - \frac{p_o b^2}{b^2 - a^2}\left(1 + \frac{a^2}{r^2}\right)$$

$$(3.72)$$

If we consider only very long cylinders, we can assume a state of *plane strain* along the cylinder's axis (normal to the plane of the paper). Then the radial displacement $u(r)$ can be determined from

$$u(r) = r\varepsilon_{\theta\theta} = \frac{(1+v)((1-v)\sigma_{\theta\theta} - v\sigma_{rr})}{E}r \qquad (3.73)$$

We can then calculate the radial displacements for the two separate cases of internal pressure (only) and external pressure (only) by substituting appropriate portions of Equation (3.72) into Equation (3.73). We then find that

$$u_i(r) = \frac{(1+v)p_i a^2}{E(b^2 - a^2)}\left[(1-2v)r + \frac{b^2}{r}\right] \qquad (3.74)$$

and

$$u_o(r) = -\frac{(1+v)p_o b^2}{E(b^2 - a^2)}\left[(1-2v)r + \frac{a^2}{r}\right] \qquad (3.75)$$

Note the different signs of these results, which suggest a real opportunity to explore our intuition.

Now, all we need to compare the radial displacements at the inner and outer surfaces is to evaluate Equation (3.74) or Equation (3.75) at $r = a$ and $r = b$, respectively, and then construct an appropriate ratio. Thus, for the case of *internal pressure* (only),

$$R_i = \frac{u_i(b)}{u_i(a)} = \left(\frac{\Delta(b)}{\Delta(a)}\right)_i = \frac{2(1-v)\rho}{(1-2v)+\rho^2} \qquad (3.76)$$

where $\rho = b/a$ is the ratio of the cylinder's outer radius to its inner radius, which perforce is such that $\rho \geq 1$. The relative size of the two displacements can be gauged according to whether the ratio R_i exceeds unity or not. The boundary for that measure is

$$\rho^2 - 2(1-v)\rho + (1-2v) = 0 \tag{3.77a}$$

which has two roots,

$$\rho_{1,2} = (1-2v), 1 \tag{3.77b}$$

The first root has no physical significance since the physics of the problem require that $\rho > 1$. Further, since $(\partial R_i/\partial\rho)_{\rho=1} = -(v/(1-v)) < 0$, it follows that the ratio R_i always decreases with ρ, and hence $u_i(b) < u_i(a)$; that is, the displacement at the outer radius is always smaller than that at the shell's inner, loaded radius.

And for the case of *external pressure* (only), we find the ratio of the inner to the outer radial displacements to be

$$R_o = \frac{u_o(a)}{u_o(b)} = \left(\frac{\Delta(a)}{\Delta(b)}\right)_o = \frac{2(1-v)\rho}{1+(1-2v)\rho^2} \tag{3.78}$$

The assessment of the magnitude of the ratio R_o with ρ is not quite as straightforward as the previous case. The relative size of the two displacements can again be gauged according to how the ratio R_o compares to unity. In this case the boundary for that measure is

$$(1-2v)\rho^2 - 2(1-v)\rho + 1 = 0 \tag{3.79a}$$

which has two physically tenable roots:

$$\rho_{1,2} = 1, 1/(1-2v) \tag{3.79b}$$

Then, since $(\partial R_o/\partial\rho)_{\rho=1} = -(v/(1-v)) > 0$, $R_o > 1$ for thinner cylinders for which $1 < \rho < 1/(1-2v)$, and $R_o < 1$ for thicker cylinders for which $1/(1-2v) < \rho$. Further, R_o reaches a peak value of 1.107 when $\rho = \rho_{cr} = 1/\sqrt{(1-2v)} = 1.581$ ($v = 0.30$). Thus, the displacement at the inner radius is always greater than that at the shell's outer, loaded radius for relatively thin cylinders, but it is always smaller than that at the outer radius for thicker cylinders. Indeed, the analysis clearly shows that as the outer radius increases indefinitely, the displacement at the inner radius tends to zero. In other words, a pressure applied at an indefinitely large radius does not affect the inner radius at all.

By way of contrast, we now turn to thin disks, for which we can assume a state of *plane stress* through the disk's axis. Here the radial displacement $u(r)$ is determined by

$$u(r) = r\varepsilon_{\theta\theta} = \frac{(\sigma_{\theta\theta} - v\sigma_{rr})}{E} r \qquad (3.80)$$

We again calculate radial displacements for the two separate cases of internal (only) and external (only) pressures by substituting appropriate portions of Equation (3.72) into Equation (3.80). We then find that

$$u_i(r) = \frac{p_i a^2}{E(b^2 - a^2)}\left[(1-v)r + (1+v)\frac{b^2}{r}\right] \qquad (3.81)$$

and

$$u_o(r) = -\frac{p_o b^2}{E(b^2 - a^2)}\left[(1-v)r + (1+v)\frac{a^2}{r}\right] \qquad (3.82)$$

We once again see different signs in these results.

Now, to compare the radial displacements at the inner and outer surfaces, we again evaluate Equations (3.81) and (3.82) at $r = a$ and $r = b$, respectively. For the case of *internal pressure* (only), the ratio of outer to inner displacements is

$$R_i = \frac{u_i(b)}{u_i(a)} = \left(\frac{\Delta(b)}{\Delta(a)}\right)_i = \frac{2\rho}{(1-v)+(1+v)\rho^2} \qquad (3.83)$$

We can again gauge the relative size of the two displacements by asking whether the ratio R_i exceeds unity or not. The boundary for that measure is

$$(1+v)\rho^2 - 2\rho + (1-v) = 0 \qquad (3.84a)$$

which has the two roots,

$$\rho_{1,2} = \frac{1-v}{1+v}, 1 \qquad (3.84b)$$

The first root has no physical significance since the physics of the problem require that $\rho > 1$. Further, since $(\partial R_i/\partial\rho)_{\rho=1} = -v < 0$, it follows that the ratio R_i always decreases with ρ, and hence $u_i(b) < u_i(a)$; that is, the displacement at the outer radius is always smaller than that at the shell's inner, loaded radius.

Finally, for *external pressure* (only) the ratio of inner-to-outer radial displacements is

$$R_o = \frac{u_o(a)}{u_o(b)} = \left(\frac{\Delta(a)}{\Delta(b)}\right)_o = \frac{2\rho}{(1+v)+(1-v)\rho^2} \qquad (3.85)$$

The assessment of the magnitude of the ratio R_o with ρ is not quite as straightforward as the previous case. The relative size of the two displacements can again be gauged according to how the ratio R_o compares to unity. In this case the boundary for that measure is

$$(1-v)\rho^2 - 2\rho + (1+v) = 0 \qquad (3.86a)$$

which has two physically tenable roots,

$$\rho_{1,2} = 1, \frac{1+v}{1-v} \qquad (3.86b)$$

We can readily see that so there is a region $1 \le \rho \le \sqrt{(1+v)/(1-v)}$ for which $R_o(\rho) \ge 1$ because $(\partial R_o/\partial \rho)_{\rho=1} = v > 0$. The displacement ratio reaches a maximum value $R_o(\rho = \sqrt{(1+v)/(1-v)} = 1/\sqrt{1-v^2}$, after which it decreases in value.

Conclusions

We have devoted this chapter to analyzing several simple models to allow us to elicit some estimates of structural behavior, with the aim of seeing whether those estimates confirm or confound our intuition. Conversely, by undertaking these analyses in a style in which we emphasized the role of dimensions and dimensional reasoning, we were able to show how we might develop intuitions about various structures. Thus, we looked at the relative magnitudes of stresses and displacements as a function of a beam's geometry and the ways in which concentrated loads could be superposed to develop uniform loads and we saw how similar reasoning allowed us to see that a simple truss is basically little more than an articulated beam. We also calculated and presented the displacements for elastic cylinders and disks under internal and external pressures, and we found that predictions of their relative magnitudes are not easily made. Thus, all of these analyses should prove useful for students, teachers, and practitioners to test and improve their physical intuition.

Bibliography

Cook, R. D., and W. C. Young. 1999. *Advanced mechanics of materials,* 2nd ed. Upper Saddle River, NJ: Prentice Hall.

Dym, C. L. 1997. *Structural modeling and analysis.* New York: Cambridge University Press.

Dym, C. L., and I. H. Shames. 1973. *Solid mechanics: A variational approach.* New York: McGraw-Hill.

Dym, C. L., and H. E. Williams. 2007. Exploring physical intuition in elementary pressure vessels. *International Journal of Mechanical Engineering Education* 35 (2): 108–113.

Fung, Y. C. 1965. *Foundations of solid mechanics.* Englewood Cliffs, NJ: Prentice Hall.

Gere, J. M., and S. P. Timoshenko. 1997. *Mechanics of materials,* 4th ed. Boston: PWS Publishing Company.

Hanson, J. H., and P. D. Brophy. 2008. Preliminary results from teaching students how to evaluate the reasonableness of results. *Proceedings of the 2008 ASEE IL/IN Section Conference: Engineering Education at the Crossroads,* Rose-Hulman Institute of Technology, Terre Haute, IN: <http://ilin.asee.org/Conference2008/Author.html> (accessed May 13, 2011).

Jolley, L. B. W. 1961. *Summation of series.* New York: Dover Publications.

Timoshenko, S. P., and J. N. Goodier. 1934. *Theory of elasticity,* 3rd ed. New York: McGraw-Hill.

Problems

3.1 Calculate the moment $M(x)$ and shear $V(x)$ resultants for a tip-loaded elementary cantilever and use them to calculate the respective stored energies due to bending and to shear. Compare the value of the two. (*Hint:* Remember how strain and complementary energies relate for linear elastic systems.)

3.2 Calculate the moment $M(x)$ and shear $V(x)$ resultants for a centrally loaded simply supported beam and use them to calculate the respective stored energies due to bending and to shear. Compare the value of the two. (*Hint:* Remember how strain and complementary energies relate for linear elastic systems.)

3.3 Using either Castigliano's second theorem or superposition (or any other technique!), calculate the tip deflection of a cantilever beam subject to three loads $P/3$ spaced at the beam's third points.

3.4 Using either Castigliano's second theorem or superposition (or any other technique!), calculate the tip deflection of a cantilever beam subject to four loads $P/4$ spaced at the beam's quarter points.

3.5 Using either Castigliano's second theorem or superposition (or any other technique!), calculate the tip deflection of a cantilever beam subject to n loads P/n spaced at uniform intervals L/n measured from the tip. Sum the resulting series for the tip deflection and compare that with the tip deflection under a uniform load q_0.

3.6   For the truss shown here, assuming $n = 2$, find the deflection of the joint at which the load F is applied.

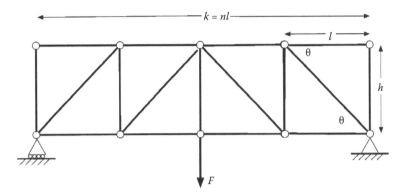

3.7   For the truss shown in Problem 3.6, assuming $n = 4$, find the deflection of the joint at which the load F is applied.

3.8   For the truss shown in Problem 3.6, for arbitrary even values of n, find the deflection of the joint at which the load F is applied. What happens to the deflection under F as n gets very large?

3.9   Define the wall thickness h of a cylinder as the difference between the outer and inner radii, $h = b - a$, and introduce a mean radius R of the cylinder such that $R = (a + b)/2$. Now transform the radial coordinate such that $r = R + z$ and rewrite the equilibrium Equation (3.71) in terms of the new variable z. How does the equilibrium equation change if h is small?

3.10   With the new radial coordinate $r = R + z$ introduced in Problem 3.9, rewrite the stresses (Equation 3.72) in terms of the new variable z. How do these two stresses change when h is small? Does this result change the analysis of the equilibrium equation of Problem 3.9?

3.11   With the new radial coordinate $r = R + z$ introduced in Problem 3.9, rewrite the displacements at the inner and outer cylinder surfaces, Equations (3.74) and (3.75), in terms of the new variable z. How do these two displacements change for small values of h (i.e., $h/R \ll 1$)? What does this imply as a possible idealization for relatively thin cylinders?

3.12   Problems 3.9–3.11 were done for long cylinders in a state of (axial) plane strain. Would the results be different (or not) were these problems done for the case of axial plane stress?

4

Estimating Stresses and Displacements in Arches

Summary

In this chapter we develop analytical estimates of the behavior exhibited by curved, arch-like structures under centrally directed and gravitational line loads. We will see that the behavior ranges from elementary beam bending at one end to a state of pure compression at the other, and that we can track that behavior through an arch rise parameter that is a function of the arch's semivertex angle, radius, and thickness. Our principal results are useful estimates of the dependence of the major displacements and stress resultants on the arch rise parameter. We will also show that these results are obtained for three different curvature expressions, and we will delineate some differences between curved beam and arch theories. Finally, we will offer some insight into the assumptions underlying the very famous arch designs of the Swiss architect Robert Maillart.

From a modeling perspective, we aim to show in this chapter (1) how assumptions about magnitude, in this case the assumption of small angles, translate into neat, useful estimates; and (2) how dimensional considerations can increase our understanding of the behavior captured in analytical models.

Introduction

Historically, "true" arches are masonry structures such as Roman bridges and arches whose basic elements are the stones of which the arch is constructed. The dominant, and perhaps only, stress state is one of uniform compression. On the other hand, structures such as Robert Maillart's beautiful bridges are perhaps better characterized as "arch-like" structures whose static response is governed not only by their extensional stiffness ($\sim EA/R$), but also by their

FIGURE 4.1
Robert Maillart's Salginatobel Bridge, Schiers, Switzerland. (Photo by J. Blumer-Maillart; with permission of Annabella Schälchli, granddaughter, and Annabella Schälchli-Cuniberti, great-granddaughter, and ASCE.)

bending stiffness ($\sim EI/R^3$). The engineer-cum-historian David Billington has described Maillart's famed Salginatobel Bridge (Figure 4.1) as "one of the most beautiful examples of pure twentieth-century structure...even to the skilled engineer, an object of mystery and wonder." He further noted that Maillart's "bridge calculations employed elementary mathematics with no calculus at all" to produce a form that without precedent expressed "one of the simplest of all technical ideas": a simple arch acting in pure compression. This elegant structural concept formed the basis of Maillart's work, as well of the bridges designed by many others (for example, Christian Menn).

Indeed, while it is "well known" that beams bend and that arches act largely in compression, it is not always obvious how an arch structure behaves or which stresses are dominant, especially to students and engineers inexperienced in the design of bridges or thin shell structures. Neither the standard structural mechanics texts nor the structural engineering professional literature tells us very much about how arch behavior changes as arch geometry varies. In fact, our review of the standard structures textbooks failed to produce even a single discussion of how or when a shallow curved beam transitions to an arch, or—and perhaps more importantly from a modeling point of view—why arch analyses are routinely conducted in terms of the arch's bending stiffness (i.e., EI/R^3), when such arches are presumed to be in compression and thus governed by their extensional stiffness (i.e., EA/R).

We want to provide some guidance on that very point. We start by examining the linear response of circular arches and arch-like structures to radial and tangential line loading for both pinned and clamped supports. Our point is to model the linear response of solid arches (i.e., without interior hinges), standing on their own or perhaps serving as the "foundations" of deck-stiffened arches. We consider both centrally applied and gravitational loading, and we will see that arch behavior ranges from pure compression to the bending stresses and deflections of elementary beams. That behavior change can be explicitly tracked by an arch rise parameter, λ, that is a function of the arch's semivertex angle α, thickness h, and radius R.

We are concerned with rather thin arches whose thickness-to-radius ratio h/R is small compared to one, so we assume that stresses vary linearly through the thickness. However, we will discuss three different curvature expressions that are used to model the bending strain, and we will see that distinctions between arch and beam theories stem entirely from a common expression for the tangential strain at the arch's midsurface. Two of the three curvature expressions used derive from the well-known Donnell and Sanders shell theories, while the third—shallow arch theory—was developed and used to model the stability behavior of very shallow arches. As we said before, we find virtually identical responses from all these theories, which confirms a long-held view that the curvature differences are very small: of the order of ε_0/R, where ε_0 is the midsurface extensional strain. Thus, the estimates of the important displacement and stress resultants that we develop later can be applied with confidence to such arch structures over wide ranges of the arch opening angle and the arch rise parameter.

Normal and Bending Stresses in Transversely Loaded Arches

Consider first a planar circular arch of constant mean radius R and thickness h, loaded on its outer surface by a radially directed (positive inward) line load q_r and a tangential line load q_θ (namely, Figure 4.2). Our starting point for modeling of the response of the arch to the load is planar strain–displacement relations of two-dimensional elasticity theory written in polar

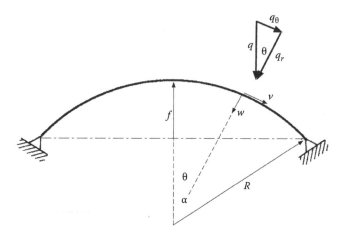

FIGURE 4.2
Geometry of and loading on a thin circular (pinned) arch of radius R and semivertex angle α.
(By permission of ASCE.)

coordinates r and θ, very much like our discussion of pressurized cylinders in Chapter 3. Those kinematic relations are

$$\varepsilon_{rr}(r,\theta) = \frac{\partial u_r(r,\theta)}{\partial r}$$

$$\varepsilon_{\theta\theta}(r,\theta) = \frac{u_r(r,\theta)}{r} + \frac{1}{r}\frac{\partial u_\theta(r,\theta)}{\partial \theta} \qquad (4.1)$$

$$\gamma_{r\theta}(r,\theta) = +\frac{1}{r}\frac{\partial u_r(r,\theta)}{\partial \theta} + \frac{\partial u_\theta(r,\theta)}{\partial r} - \frac{u_\theta(r,\theta)}{r}$$

where $u_r(r,\theta)$ and $u_\theta(r,\theta)$ are, respectively, the (outward) radial and tangential displacements of a point in the plane of the arch.

We now replace the radial coordinate with a thickness coordinate z measured positive inward from the arch's centroidal radius R according to $r = R - z$, and we approximate the radial displacement as positive inward and uniform through the thickness (namely, Figure 4.3),

$$u_r(r,\theta) = -w(R,\theta) \equiv -w(\theta) \qquad (4.2a)$$

and the tangential displacement as a linear function of the thickness coordinate,

$$u_\theta(r,\theta) = v(\theta) + \frac{z}{R}v_1(\theta) \qquad (4.2b)$$

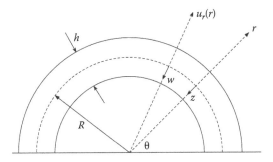

FIGURE 4.3
A representation of circular and shell coordinates, r and $R - z$, respectively, and the corresponding radial displacement $u_r(r) = -w(z)$.

The displacement formulations (Equations 4.2a and 4.2b) are one part of the standard Euler–Bernoulli assumptions we use to model the distribution of strain and displacement through the thickness of relatively thin structural elements. The other significant Euler–Bernoulli assumption we make is that the shear *strain*—not the shear *stress*: (recall our discussion of beams in Chapter 3!)—vanishes through the thickness, from which it follows that

$$v_1(\theta) \cong -\left(v(\theta) + \frac{dw(\theta)}{d\theta} \right) \tag{4.3}$$

It then also follows from Equations (4.2a), (4.2b), and (4.3) that $\varepsilon_{rr}(r, \theta) = 0$ and

$$\varepsilon_{\theta\theta}(z, \theta) \cong \left(\frac{1}{R-z} \right) \left[\left(\frac{dv(\theta)}{d\theta} - w(\theta) \right) - \frac{z}{R} \left(\frac{d^2 w(\theta)}{d\theta^2} + \frac{dv(\theta)}{d\theta} \right) \right] \tag{4.4}$$

Equation (4.4) contains within it the ring version of two well-known approximations for the behavior of thin circular shells. The first is a linear subset of a nonlinear ring equation that is itself a one-dimensional subset of the famous Sanders nonlinear shell equations that are regarded as the "gold standard" of cylindrical shell equations. The Sanders approximation is found here simply by assuming that $R - z \cong R$, in which case Equation (4.4) becomes

$$\varepsilon_{\theta\theta}(z, \theta) \cong \left[\frac{1}{R} \left(\frac{dv(\theta)}{d\theta} - w(\theta) \right) - \frac{z}{R^2} \left(\frac{d^2 w(\theta)}{d\theta^2} + \frac{dv(\theta)}{d\theta} \right) \right] \tag{4.5}$$

The second approximation is known as the Donnell approximation and is found by expanding the term $R - z$ in Equation (4.4) to first-order powers of the thickness coordinate z—that is,

$$\varepsilon_{\theta\theta D}(z,\theta) \cong \left(1 + \frac{z}{R}\right)\left[\left(\frac{v(\theta)}{d\theta} - w(\theta)\right) - \frac{z}{R}\left(\frac{d^2w(\theta)}{d\theta^2} + \frac{dv(\theta)}{d\theta}\right)\right]$$

$$\cong \left[\frac{1}{R}\left(\frac{v(\theta)}{d\theta} - w(\theta)\right) - \frac{z}{R^2}\left(w(\theta) + \frac{d^2w(\theta)}{d\theta^2}\right)\right]$$

(4.6)

The difference between the Sanders approximation (Equation 4.5) and its Donnell counterpart (Equation 4.6) is that their respective curvatures terms differ by the term

$$\frac{1}{R^2}\left(\frac{d^2w(\theta)}{d\theta^2} + w(\theta)\right) = \frac{1}{R^2}\left(\frac{d^2w(\theta)}{d\theta^2} + \frac{dv(\theta)}{d\theta}\right) + \frac{1}{R^2}\left(w(\theta) - \frac{dv(\theta)}{d\theta}\right) \quad (4.7)$$

We will later show that this difference is small and that it makes no discernible difference in the arch behaviors we predict with these two approximations (as well as with a third such approximation).

We will use the strain–displacement relation (Equation 4.5) as our benchmarking gold standard. With it we can write two constitutive relations to define axial and bending stress resultants, $N(\theta)$ and $M(\theta)$, in terms of strains and displacements consistent with Equation (4.5):

$$N(\theta) = \frac{EA}{R}\left(\frac{dv(\theta)}{d\theta} - w(\theta)\right)$$

$$M(\theta) = -\frac{EI}{R^2}\left(\frac{d^2w(\theta)}{d\theta^2} + \frac{dv(\theta)}{d\theta}\right)$$

(4.8)

where E is Young's modulus and A and I are, respectively, the arch's cross-sectional area and the area's second moment.

For an arch subject to forces per unit length q_r and q_θ, we can write the first variation of the total potential energy corresponding to the kinematic result (Equation 4.5) as

$$\delta^{(1)}\Pi = \int_{-\alpha}^{\alpha}\left\{\frac{N}{R}\left[\frac{d\delta v}{d\theta} - \delta w\right] - \frac{M}{R^2}\left(\frac{d^2\delta w}{d\theta^2} + \frac{d\delta v}{d\theta}\right) - q_\theta\delta v - q_r\delta w\right\}R\,d\theta \quad (4.9)$$

where α is the arch's semivertex angle; that is, 2α is the arch opening angle. Also, consistent with the various kinematic formulations used herein, we are assuming that the arch is thin so that $(R + z)d\theta \cong Rd\theta$ in formulating the integral (Equation 4.9). If we set the first variation of the total potential (Equation 4.9) to zero, we find two equations of equilibrium for the loaded arch:

$$\frac{1}{R}\frac{dN(\theta)}{d\theta} - \frac{1}{R^2}\frac{dM(\theta)}{d\theta} + q_\theta = 0$$
$$\frac{N(\theta)}{R} + \frac{1}{R^2}\frac{d^2M(\theta)}{d\theta^2} + q_r = 0 \tag{4.10}$$

If we introduce a transverse shear stress resultant, very much as in beam theory,

$$Q(\theta) = \frac{1}{R}\frac{dM(\theta)}{d\theta} \tag{4.11}$$

we find that the equations of equilibrium become

$$\frac{dN(\theta)}{d\theta} - Q(\theta) + q_\theta R = 0$$
$$N(\theta) + \frac{dQ(\theta)}{d\theta} + q_r R = 0 \tag{4.12}$$

When we set the first variation (Equation 4.9) of the total potential to zero, we also obtain consistent choices of boundary conditions at the ends of the arch, $\theta = \pm\alpha$—namely:

$$\text{Either} \quad N(\pm\alpha) - \frac{M(\pm\alpha)}{R} = 0 \quad \text{or} \quad \delta v(\pm\alpha) = 0 \tag{4.13a}$$

$$\text{Either} \quad Q(\pm\alpha) = \frac{1}{R}\frac{dM(\pm\alpha)}{d\theta} = 0 \quad \text{or} \quad \delta w(\pm\alpha) = 0 \tag{4.13b}$$

$$\text{Either} \quad M(\pm\alpha) = 0 \quad \text{or} \quad \delta\left(\frac{dw(\pm\alpha)}{d\theta}\right) = 0 \tag{4.13c}$$

We now want to make one more point before solving the equations we just derived. Those equations are valid for any relatively thin arch (or ring) for which $h/R \ll 1$, but for any value of the circumferential angle θ. However, it will prove useful to make an additional modeling assumption—namely, that all angles θ are sufficiently small that $\sin\theta \cong \theta$

and $\cos\theta \cong 1$ or $\cos\theta \cong 1 - \theta^2/2$, depending on the context. Note that this "small angle" approximation is good even for relatively large angles: For $\alpha = 45° \cong 0.79$ rad, $\sin(0.79) = \cos(0.79) \cong 0.71$ and $1 - (0.71)^2/2 \cong 0.69$. Thus, the small angle approximation can include arches whose total opening angles are as large as $2\alpha = 90° \cong 1.572$ rad. The physical import of the small angle assumption is that the arch rise f (see Figure 4.2) can be written and then approximated as

$$\frac{f}{R} = (1 - \cos\alpha) \cong \frac{\alpha^2}{2} \qquad (4.14)$$

Thus, the assumption of small angles implies a concomitant assumption that the arch has a small rise and may be considered a shallow arch.

Arches under Centrally Applied, "Dead" Loading

Assuming that the arch responds symmetrically to a (symmetric) uniform line load $q_r = q$ with $q_\theta = 0$, we can find the stress resultants by integrating Equations (4.10) and (4.11):

$$N(\theta) = -qR(1 + C_1 \cos\theta)$$

$$M(\theta) = qR^2 (C_2 - C_1 \cos\theta) \qquad (4.15)$$

$$Q(\theta) = qR(C_1 \sin\theta)$$

We find the constants C_i by satisfying stress boundary conditions, after which we formulate equations for the displacements v and w by substituting the stress resultants (Equation 4.15) into the constitutive Equation (4.8). After some manipulation, we find the following governing equations for the displacements:

$$\frac{dv(\theta)}{d\theta} - w(\theta) = -\frac{qR^2}{EA}(1 + C_1 \cos\theta)$$

$$\frac{d^2w(\theta)}{d\theta^2} + \frac{dv(\theta)}{d\theta} = -\left(\frac{qR^2}{EA}\right)\left(\frac{1}{\bar{I}}\right)(C_1 \cos\theta - C_2) \qquad (4.16)$$

Note that we have introduced a dimensionless ratio $\bar{I} \triangleq I / AR^2$ into Equation (4.16); for rectangular cross sections ($A = bh$, $I = bh^3/12$), this means that $\bar{I} = h^2 / 12R^2$. Equation (4.16) is readily uncoupled and integrated, enabling us to find the (inwardly positive) radial displacement (which is symmetric about $\theta = 0$) as

$$w(\theta) = \frac{qR^4}{EI}\left[\left(\left(\frac{1+\bar{I}}{2}\right)\theta \sin\theta\right)C_1 - C_2 + \left(\bar{I}\cos\theta\right)C_3 + \bar{I}\right] \tag{4.17}$$

and the tangential displacement (which is antisymmetric about $\theta = 0$) as

$$v(\theta) = \frac{qR^4}{EI}\left[\left((1-\bar{I})\sin\theta - (1+\bar{I})\theta\cos\theta\right)\frac{C_1}{2} - (\bar{I}\theta)C_2 + (\bar{I}\sin\theta)C_3\right] \tag{4.18}$$

We now determine the unknown constants (C_1, C_2, C_3) in the solutions (Equations 4.15, 4.17, and 4.18) for a pinned (simply supported) arch, for which the boundary conditions are

$$w(\pm\alpha) = 0, \quad v(\pm\alpha) = 0, \quad M(\pm\alpha) = 0 \tag{4.19}$$

When we satisfy the boundary conditions (Equation 4.19), we find three equations for the three unknown constants; their solutions for the pinned arch are

$$C_{1p} = -\frac{2\bar{I}\sin\alpha}{D_p}, \quad C_{2p} = -\frac{\bar{I}\sin 2\alpha}{D_p}, \quad C_{3p} = \frac{(\sin\alpha - 3\alpha\cos\alpha) - \bar{I}(\sin\alpha + \alpha\cos\alpha)}{D_p} \tag{4.20}$$

where the common denominator D_p is

$$D_p = (2\alpha + \alpha\cos 2\alpha - (3/2)\sin 2\alpha) + \bar{I}(\alpha + \sin 2\alpha/2) \tag{4.21}$$

When we replace the trigonometric terms in Equation (4.21) by their small angle approximations, we find that

$$D_p \cong 2\alpha\bar{I}(1 + 2\alpha^5/(15\bar{I})) \tag{4.22}$$

Now, given that both α and $\bar{I}$ are small numbers, how do we evaluate the dimensionless factor $2\alpha^4/15\bar{I}$ in Equation (4.22)? The very structure of that equation suggests that α and $\bar{I}$ are not independent and that the factor $2\alpha^4/15\bar{I}$ is a parameter that needs to be considered. In fact, the factor $2\alpha^4/15\bar{I}$ is a variant of an arch rise parameter that was first introduced to analyze the postbuckling behavior of shallow arches. In that tradition, we define a dimensionless *arch rise parameter* λ as the ratio of the arch rise f to one-half of the arch thickness h:

$$\lambda \triangleq \frac{f}{h/2} = \frac{2R}{h}(1 - \cos\alpha) \cong \alpha^2 \frac{R}{h} \tag{4.23}$$

Given the definitions of $\bar{I}$ and λ, we see that

$$\bar{I} = I/AR^2 = \alpha^4/(12\lambda^2) \tag{4.24}$$

Thus, we can recast Equation (4.22) as

$$D_p \cong 2\alpha\bar{I}(1 + 8\lambda^2/5) \tag{4.25}$$

We can similarly find corresponding approximations for the constants C_{ip} (Equation 4.20), but it is important to remember that we rarely evaluate such constants in isolation. For example, we would find the moment resultant at the crown ($\theta = 0$) of a pinned arch from Equation (4.15) as

$$M(0) = qR^2(C_{2p} - C_{1p}) = qR^2 \frac{\bar{I}(2\sin\alpha - \sin 2\alpha)}{D_p}$$

$$\cong qR^2 \frac{\bar{I}\left[2(\alpha - \alpha^3/6) - (2\alpha - 4\alpha^3/3)\right]}{D_p} = \frac{qR^2\bar{I}\alpha^3}{D_p} \tag{4.26}$$

We see here that we must keep higher order terms in α^n (sometimes to $n = 5$ or 6) in order to avoid errors by prematurely truncating our series expansions.

We also note in this context that the developed length of a shallow arch is essentially equal to its span length, $L = 2R\sin\alpha \cong 2R\alpha = S$, due to our small angle assumption. With this in mind and with the aid of Equation (4.25), we can write Equation (4.26) as

$$M(0) \cong \frac{q(R\alpha)^2 \bar{I}\alpha}{2\alpha\bar{I}(1 + 8\lambda^2/5)} = \frac{qL^2/8}{(1 + 8\lambda^2/5)} \tag{4.27}$$

The preceding analysis suggests that for small angles α, arch behavior depends explicitly on the arch rise parameter λ, and that λ will be influential in delineating different regimes of arch behavior. In particular, we will later see that for small values of λ ($\lambda \ll 1$), the arch in fact flattens completely and behaves like a beam in bending; for large values of λ ($\lambda \gg 1$), the arch behaves as a true arch with a uniformly compressive state of stress. Note that in each of the preceding inequalities, we are using 1 as our standard or unit of measurement. In fact, we will show that our standard will depend on the boundary conditions: for pinned arches our standard will compare $8\lambda^2/5$ to 1 (as suggested in Equation 4.27), while for clamped arches we will compare $4\lambda^2/15$ to 1. Interestingly, we will also see these same two factors when we formulate the same problems using two other arch curvatures.

In principle, we have now determined the two displacement components and the three corresponding stress resultants for simply supported arches. However, because we are primarily interested in shallow arches, we will simplify our general solutions (Equations 4.15, 4.17, and 4.18) by expanding the trigonometric terms in the solutions to account both for small semivertex angles α and for correspondingly small values of the circumferential coordinate θ, since $-\alpha \le \theta \le \alpha$. (We will present some numerical values and limiting cases later in this chapter.) For pinned arches, with particular attention to orders of magnitude, it turns out that the (inward) radial deflection at the crown of the arch ($\theta = 0$) is

$$w^p(0) \cong \frac{5\lambda^2/2}{1+8\lambda^2/5}\left(\frac{qR^2}{EA}\right) \cong \frac{1}{1+8\lambda^2/5}\left(\frac{5qL^4}{384EI}\right) \tag{4.28}$$

The axial stress resultant behaves as

$$N^p(\theta) \cong -qR\left(\frac{8\lambda^2/5}{1+8\lambda^2/5}\right) \tag{4.29}$$

Similarly, the moment resultant is

$$M^p(\theta) \cong \frac{qL^2/8}{1+8\lambda^2/5}(1-(\theta/\alpha)^2) \tag{4.30}$$

The transverse shear is

$$Q^p(\theta) \cong -\frac{qL/2}{1+8\lambda^2/5}(\theta/\alpha) \tag{4.31}$$

We now examine the arch's circumferential stress. First, we write it in terms of the stress resultants (Equation 4.8):

$$\sigma_{\theta\theta}(z,\theta) = \frac{N(\theta)}{A} + \frac{M(\theta)z}{I} \tag{4.32}$$

Then we evaluate that stress at the crown of the pinned arch by substituting the in-plane (Equation 4.29) and bending (Equation 4.30) resultants, in their simple limit forms, into Equation (4.32) and, again assuming a rectangular cross section,

$$\sigma^p_{crown} = \sigma^p_{\theta\theta}(z,\theta=0) = -\frac{qR/A}{1+8\lambda^2/5}\left[\overbrace{8\lambda^2/5}^{compression} - \left(\frac{2z}{h}\right)\overbrace{3\lambda}^{bending}\right] \tag{4.33}$$

The last pinned arch results we will determine are the horizontal (inwardly directed) and vertical (upwardly directed) forces at the arch supports. Maintaining our small angle assumption and with the moment at the supports ($\theta = \pm\alpha$) clearly zero, we find that

$$R^p_H = qR\left(\frac{(8\lambda^2/5)}{1+8\lambda^2/5}\right) \quad \text{and} \quad R^p_V = qR\alpha = \frac{qL}{2} \tag{4.34}$$

Note that the arguments of all the results we have given in Equations (4.28)–(4.31), (4.33), and (4.34) for the pinned arch depend explicitly only on the arch rise parameter λ.

For an arch that is clamped at its boundaries, the appropriate boundary conditions are

$$w^c(\pm\alpha)=0, \quad v^c(\pm\alpha)=0, \quad \frac{dw^c(\pm\alpha)}{d\alpha}=0 \tag{4.35}$$

We now determine the unknown constants in the solutions (Equations 4.15, 4.17, and 4.18) for a clamped arch, finding that

$$C_{1c}=-\frac{2\bar{I}\alpha\sin\alpha}{D_c}, \quad C_{2c}=-\frac{2\bar{I}\sin^2\alpha}{D_c}, \quad C_{3c}=-\frac{(1+\bar{I})(\alpha\sin\alpha+\alpha^2\cos\alpha)}{D_c} \tag{4.36}$$

and the common denominator D_c is

$$D_c = (\alpha^2+\alpha\sin\alpha\cos\alpha+\cos 2\alpha-1)+\bar{I}(\alpha^2+\alpha\sin\alpha\cos\alpha) \tag{4.37}$$

As we did before for the denominator D_p, we expand the denominator (Equation 4.37) for small angles α, with the following result:

$$D_c \cong 2\alpha^2\bar{I}(1+\alpha^4/45) \equiv 2\alpha^2\bar{I}(1+4\lambda^2/15) \tag{4.38}$$

Thus, we now know the displacement components and stress resultants for clamped arches. For fixed arches for which $\alpha^2 \ll 1$, the (inward) radial deflection at the crown of the arch ($\theta = 0$) is

$$w^c(0) \cong \frac{\lambda^2/2}{1+4\lambda^2/15}\left(\frac{qR^2}{EA}\right) \equiv \frac{1}{1+4\lambda^2/15}\left(\frac{qL^4}{384EI}\right) \tag{4.39}$$

The axial stress result is

$$N^c(0) \cong -qR\left(\frac{4\lambda^2/15}{1+4\lambda^2/15}\right) \tag{4.40}$$

Similarly, the moment resultant is

$$M^c(0) \cong \frac{qL^2/24}{1+4\lambda^2/15}(1-3(\theta/\alpha)^2) \tag{4.41}$$

and the shear resultant is

$$Q^c(0) \cong -\frac{qL/2}{1+4\lambda^2/15}\left(\frac{\theta}{\alpha}\right) \tag{4.42}$$

We again find the stress at the crown of the clamped arch by substituting the appropriate stress resultants $N(\theta)$ and $M(\theta)$ into Equation (4.32), assuming as usual a rectangular cross section:

$$\sigma^c_{crown} = \sigma^c_{\theta\theta}(z, \theta = 0) = -\frac{qR/A}{1+4\lambda^2/15}\left[\overbrace{4\lambda^2/15}^{compression} - \left(\frac{2z}{h}\right)\overbrace{\lambda}^{bending}\right] \tag{4.43}$$

Our final results for the clamped arch are the (inward) horizontal and (upward) vertical support forces and the support moment, which for small angles are

$$R^c_H = qR\left(\frac{4\lambda^2/15}{1+4\lambda^2/15}\right), \quad R^c_V = qR\alpha = \frac{qL}{2}, \quad M^c(\pm\alpha) = -\frac{qL^2/12}{1+4\lambda^2/15} \tag{4.44}$$

And we see for the clamped arch, as we did with the pinned arch, that the arguments of all the results given in Equations (4.39)–(4.44) depend only on the arch rise parameter λ.

What would happen if we repeated this analysis of pinned and clamped arches under centrally directed or dead pressure with the Donnell curvature formulation (Equation 4.6)? We had noted earlier that the Sanders curvature (Equation 4.5) and the Donnell curvature (Equation 4.6) differ by a very small term, which we identified in Equation (4.7) and rewrite here as

$$
\frac{1}{R^2}\left(\frac{d^2w(\theta)}{d\theta^2}+w(\theta)\right)=\frac{1}{R^2}\left(\frac{d^2w(\theta)}{d\theta^2}+\frac{dv(\theta)}{d\theta}\right)\left[1-\frac{\left(\dfrac{dv(\theta)}{d\theta}-w(\theta)\right)}{\left(\dfrac{d^2w(\theta)}{d\theta^2}+\dfrac{dv(\theta)}{d\theta}\right)}\right] \quad (4.45)
$$

The difference between the two curvatures is embedded in the ratio in Equation (4.45), so it would be very helpful if we could estimate just how big that ratio is. We make that estimate by noting that the ratio's numerator and denominator represent, respectively, the normal and bending stress resultants of Equation (4.8). Then, using our pinned arch solutions (Equations 4.29 and 4.30) to quantify those resultants at the crown ($\theta = 0$),

$$
\left.\frac{\left(\dfrac{dv(\theta)}{d\theta}-w(\theta)\right)}{\left(\dfrac{d^2w(\theta)}{d\theta^2}+\dfrac{dv(\theta)}{d\theta}\right)}\right|_{\theta=0}=\frac{\dfrac{RN(0)}{EA}}{\dfrac{R^2M(0)}{EI}}=\left(\frac{8\lambda^2}{5}\right)\left(\frac{8I}{AL^2}\right)\cong\frac{4\alpha^2}{15}\ll 1 \quad (4.46)
$$

The inequality in Equation (4.46) follows from our earlier small angle approximation.

We can see the very smallness of this difference in the curvatures when we inspect the equations of equilibrium that are consistent with Equation (4.6)—namely:

$$
\frac{dN_D(\theta)}{d\theta}+q_\theta R = 0
$$

$$
N_D(\theta)+\frac{1}{R}\frac{d^2M_D(\theta)}{d\theta^2}+\frac{M_D(\theta)}{R}+q_r R = 0
$$

$$(4.47)$$

The principal distinction between the Donnell-based equilibrium Equation (4.47) and the Sanders-based counterparts (Equation 4.10) is that here we see quite clearly that the axial resultant $N_D(\theta)$ is a constant when there is no tangential load q_θ. In fact, the particulars of a Donnell-type solution are so similar to those of our previous Sanders-based analysis that there is no noticeable difference in the end results. In other words, *all* of the results that we have presented (i.e., the circumferential stress $N(\theta)$, moment $M(\theta)$, shear $Q(\theta)$, and the deflection $\hat{w}(0)$ at the crown) are the same as those given before for both curvature assumptions for the small angle approximation and for the limiting λ cases that we will discuss in more detail later.

Shallow Arches under Centrally Directed, "Dead" Loading

We now consider a *shallow* planar circular arch of constant mean radius R and thickness h, loaded on its outer surface by a uniform, centrally directed line load q. The starting point for this model of the arch response is founded on a simplification of the curvature such that the circumferential strain is given as

$$\varepsilon_{\theta\theta}(z,\theta) = \frac{1}{R}\left(\frac{dv(\theta)}{d\theta} - w(\theta)\right) - \frac{z}{R^2}\frac{d^2w(\theta)}{d\theta^2} \tag{4.48}$$

The shallow arch strain–displacement relation (Equation 4.48) differs from its Sanders counterpart by another very small term—that is, from Equations (4.5) and (4.48),

$$\frac{1}{R^2}\left(\frac{d^2w(\theta)}{d\theta^2}\right) = \frac{1}{R^2}\left(\frac{d^2w(\theta)}{d\theta^2} + w(\theta)\right) + \frac{1}{R^2}(-w(\theta)) \tag{4.49a}$$

or

$$\frac{1}{R^2}\left(\frac{d^2w(\theta)}{d\theta^2}\right) = \frac{1}{R^2}\left(\frac{d^2w(\theta)}{d\theta^2} + w(\theta)\right)\left[1 - \frac{w(\theta)}{\left(\dfrac{d^2w(\theta)}{d\theta^2} + w(\theta)\right)}\right] \tag{4.49b}$$

The differences in the two curvatures are again embodied in the fraction in Equation (4.49b). For the pinned arch under dead pressure, we can write that fraction in terms of the radial displacement and the moment. Then, we can use the solutions (Equations 4.28 and 4.30) to verify that this curvature ratio is also quite small (again at the crown):

$$\frac{w(0)}{\left(\dfrac{d^2w(0)}{d\theta^2} + w(0)\right)} = \frac{w(0)}{\dfrac{R^2M(0)}{EI}} = \frac{5\alpha^2}{12} \ll 1 \tag{4.50}$$

Thus, the difference between the shallow arch curvature and the Donnell curvature is negligibly small, so we might as well simply rest here. However, it turns out that the shallow arch curvatures are part of a model for which neat, closed-form solutions are very easily obtained, and they also make some other inferences easier, so we will pursue a full shallow arch theory solution now.

Applying the same minimum energy method as before, we find the equations of equilibrium for the shallow arch (corresponding to Equation 4.48) to be

$$\frac{dN_s(\theta)}{d\theta} = 0$$

$$N_s(\theta) + \frac{1}{R}\frac{d^2M_s(\theta)}{d\theta^2} + qR = 0$$

(4.51)

The solutions to Equation (4.51) are

$$N_s(\theta) = -N_0, \quad \text{a constant}$$

$$M_s(\theta) = (N_0 R - qR^2)(\theta^2/2) + C_1\theta + C_2$$

(4.52)

The tangential and radial displacements are found by integrating the corresponding shallow arch constitutive equations under appropriate boundary and symmetry conditions. It is worth noting that the algebra for the shallow arch analysis is far simpler than for the Sanders and Donnell analyses. For a shallow pinned arch, the axial, moment, and shear resultants are, respectively,

$$N_s^p(\theta) = -N_0 = -qR\left[\frac{8\lambda^2/5}{1+8\lambda^2/5}\right]$$

(4.53)

$$M_s^p(\theta) = \frac{qL^2/8}{1+8\lambda^2/5}(1-(\theta/\alpha)^2)$$

(4.54)

and

$$Q_s^p(\theta) = -\frac{qL/2}{1+8\lambda^2/5}\left(\frac{\theta}{\alpha}\right)$$

(4.55)

The corresponding radial displacement is

$$w_s^p(\theta) = \frac{\lambda^2/2}{1+8\lambda^2/5}\left[5-6\left(\frac{\theta}{\alpha}\right)^2+\left(\frac{\theta}{\alpha}\right)^4\right]\left(\frac{qR^2}{EA}\right)$$

(4.56)

Not surprisingly, Equations (4.53)–(4.55) are precisely the same as their corresponding small angle results, and Equation (4.56) clearly matches the deflection pattern of a uniformly loaded (curved) beam whose value at the crown is exactly that given in Equation (4.28).

For the clamped shallow arch, the axial, moment, and shear resultants are, respectively,

$$N_s^c(\theta) = -N_0 = -qR\left[\frac{4\lambda^2/15}{1+4\lambda^2/15}\right] \tag{4.57}$$

$$M_s^c(\theta) = \frac{qL^2/24}{1+4\lambda^2/15}(1-3(\theta/\alpha)^2) \tag{4.58}$$

and

$$Q_s^c(\theta) = -\frac{qL/2}{1+4\lambda^2/15}\left(\frac{\theta}{\alpha}\right) \tag{4.59}$$

The corresponding radial displacement is

$$w_s^c(\theta) = \frac{\lambda^2/2}{1+4\lambda^2/15}\left[1-\left(\frac{\theta}{\alpha}\right)^2\right]^2\left(\frac{qR^2}{EA}\right) \tag{4.60}$$

Equations (4.57)–(4.59) are also precisely the same as their preceding small angle results, and Equation (4.60) clearly matches the deflection pattern of a uniformly loaded (curved) fixed-ended beam whose value at the crown is exactly that given in Equation (4.39).

Arches under Gravitational Loading

We now consider the important case of gravitational loading, for which $q_r = \rho g A\cos\theta$ and $q_\theta = \rho g A\sin\theta$. The solutions for the various arch theories appear much as the preceding ones did, albeit with significantly more algebra. The shallow arch equilibrium equations under gravitational loading are

$$\frac{dN_s(\theta)}{d\theta} + \rho g AR\sin\theta = 0$$

$$N_s(\theta) + \frac{1}{R}\frac{d^2M_s(\theta)}{d\theta^2} + \rho g AR\cos\theta = 0 \tag{4.61}$$

The solutions to Equation (4.61) are straightforwardly found to be

$$N_s(\theta) = C_1 + \rho g A R \cos\theta$$

$$M_s(\theta) = -\frac{C_1 R}{2}\theta^2 + C_2 + 2\rho g A R^2 \cos\theta \qquad (4.62)$$

The solutions (Equation 4.62) differ from their corresponding results (Equations 4.53 and 4.54) for centrally directed loading only by the appearance of two trigonometric terms in place of simple powers of the arch variable θ. Similarly, the deflection is

$$w_s(\theta) = \frac{C_1 R^3}{24EI}\theta^4 - \frac{C_2 R^2}{2EI}\theta^2 + C_3 + \frac{2\rho g A R^4}{EI}\cos\theta \qquad (4.63)$$

Not surprisingly, we find that the results for such gravitational loading on arches with small vertex angles α are essentially identical to those for the centrally loaded arch. Thus, we give only some highlights here. For example, for a pinned shallow arch model, we find the axial stress resultant is

$$N_s^p(\theta) = -\rho g A R\left(\frac{1+16\lambda^2/5}{1+8\lambda^2/5} - \cos\theta\right) \cong -\frac{8\lambda^2/5}{1+8\lambda^2/5}\rho g A R \qquad (4.64)$$

while the moment resultant is

$$M_s^p(\theta) = \rho g A R^2\left[2(\cos\theta - \cos\alpha) - \frac{1+16\lambda^2/5}{2(1+8\lambda^2/5)}(\theta^2-\alpha^2)\right]$$

$$\cong \frac{\rho g A(\alpha R)^2}{2(1+8\lambda^2/5)}\left(1-\left(\frac{\theta}{\alpha}\right)^2\right) = \frac{\rho g A L^2/8}{1+8\lambda^2/5}\left(1-\left(\frac{\theta}{\alpha}\right)^2\right) \qquad (4.65)$$

and the corresponding deflection at the crown of the pinned shallow arch is

$$w_s^p(0) = \frac{5\lambda^2/2}{1+8\lambda^2/5}\left(\frac{\rho g A R^2}{EA}\right) \cong \frac{1}{1+8\lambda^2/5}\left(\frac{5\rho g A L^4}{384EI}\right) \qquad (4.66)$$

Since $q = \rho g A$ here, it is evident that these results correspond exactly to those obtained for centrally directed loading (Equations 4.29, 4.30, and 4.28, respectively).

We can also obtain similar results using the Sanders curvature formulation, although we would find the algebra to be much more complicated than

any of the analyses we have presented heretofore. The governing equilibrium equations are, with $q \triangleq \rho g A$,

$$\frac{1}{R}\frac{dN(\theta)}{d\theta} - \frac{1}{R^2}\frac{dM(\theta)}{d\theta} + q\sin\theta = 0$$

$$\frac{N(\theta)}{R} + \frac{1}{R^2}\frac{d^2M(\theta)}{d\theta^2} + q\cos\theta = 0$$

(4.67)

In this case, we find the general solutions for the stress resultants to be

$$N(\theta) = qR(C_1\cos\theta - \theta\sin\theta)$$

$$M(\theta) = qR^2(C_1\cos\theta + C_2 - (\cos\theta + \theta\sin\theta))$$

(4.68)

and the radial deflection is

$$w(\theta) = \frac{qR^4}{4EI}\left[-(2(1+\bar{I})\theta\sin\theta)C_1 - 4C_2 + C_3\cos\theta + (3+\bar{I})\theta\sin\theta - (1+\bar{I})\theta^2\cos\theta\right]$$

(4.69)

The algebra involved for a Sanders-based analysis of a gravity-loaded arch
is considerably longer and messier than for the central loading, so we will
present only numerical resultants here. We will display values of the stress
resultants and the radial displacement alongside the exact and approximate
results for the arches under central loading, and we will see that they differ by no more than a few percent. That is, the circumferential and bending stress resultants and the radial displacement at the arch crown under
gravitational loading are virtually identical to those under centrally directed
loading when α is small (e.g., $\alpha = 0.20$). The moment resultant values differ
modestly when the arch opening angle and the arch rise parameter are large
(i.e., $\alpha = 0.80$, $\lambda \geq 4$).

Interpreting Arch Behavior

We summarize the principal results for pinned and clamped arches
under centrally directed loading in Table 4.1. We note that these analytical results, obtained under the small angle assumption, comprise a useful
set of closed-form estimators of the behaviors of the radial displacement
at the crown; the axial, moment, and shear stress resultants; the stress at

TABLE 4.1

Crown Displacements, Stress Resultants, Stresses, and Reactions for Small Values
of Opening Arch Angle α for Centrally Directed Loading

	Pinned arches	Clamped arches
Crown displacement	$w^p(0) \cong \dfrac{5\lambda^2/2}{1+8\lambda^2/5}\left(\dfrac{qR^2}{EA}\right)$	$w^c(0) \cong \dfrac{\lambda^2/2}{1+4\lambda^2/15}\left(\dfrac{qR^2}{EA}\right)$
Axial resultant	$N^p(\theta) = -qR\left(\dfrac{8\lambda^2/5}{1+8\lambda^2/5}\right)$	$N^c(\theta) = -qR\left(\dfrac{4\lambda^2/15}{1+4\lambda^2/15}\right)$
Moment resultant	$M^p(\theta) = \dfrac{qL^2/8}{1+8\lambda^2/5}\left(1-\left(\dfrac{\theta}{\alpha}\right)^2\right)$	$M^c(\theta) = \dfrac{qL^2/24}{1+4\lambda^2/15}\left(1-3\left(\dfrac{\theta}{\alpha}\right)^2\right)$
Shear resultant	$Q^p(\theta) = -\dfrac{qL/2}{1+8\lambda^2/5}\left(\dfrac{\theta}{\alpha}\right)$	$Q^c(\theta) = -\dfrac{qL/2}{1+4\lambda^2/15}\left(\dfrac{\theta}{\alpha}\right)$
Axial stress (pinned)	$\sigma_{crown}^p = \sigma_{\theta\theta}^p(z,\theta=0) = -\dfrac{qR/A}{1+8\lambda^2/5}\left[\overset{compression}{8\lambda^2/5} - \left(\dfrac{2z}{h}\right)\overset{bending}{3\lambda}\right]$	
Axial stress (clamped)	$\sigma_{crown}^c = \sigma_{\theta\theta}^c(z,\theta=0) = -\dfrac{qR/A}{1+4\lambda^2/15}\left[\overset{compression}{4\lambda^2/15} - \left(\dfrac{2z}{h}\right)\overset{bending}{\lambda}\right]$	
Horizontal reactions	$R_H^p = qR\left(\dfrac{8\lambda^2/5}{1+8\lambda^2/5}\right)$	$R_H^c = qR\left(\dfrac{4\lambda^2/15}{1+4\lambda^2/15}\right)$
Vertical reactions	$R_V^p = qL/2$	$R_V^c = qL/2$
Support moments	$M^p(\pm\alpha) = 0$	$M^c(\pm\alpha) = \mp\dfrac{qL^2/12}{1+4\lambda^2/15}$

the crown; and the reactions at the arch supports. These displacement and stress estimates are explicit functions of the arch rise parameter λ, but not of the semivertex angle α.

Now, despite our small angle assumption, these results are also quite similar for comparatively large values of α. As we see first in Table 4.2, which compares exact results from the Sanders formulation to its small angle approximation, the axial and moment resultants and the radial or transverse displacement are virtually identical for $\alpha = 0.20$ and $\alpha = 0.80$.

Table 4.3 shows us results for clamped arches that are almost as good, with a 33% difference in the axial resultant and a 15% difference in the radial displacement when $\lambda = 1$. The data in Tables 4.2 and 4.3 also show consistent variations in the various physical quantities as the value of the arch rise parameter λ changes, but they do not change very much with the opening angle α. The data in Tables 4.2 and 4.3 thus confirm that our estimates of the

TABLE 4.2

Exact and Approximate (Small α) Axial and
Moment Stress Resultants and Radial
Displacements at the Crown of *Pinned* Arches
under Centrally Directed and Gravity ($q = \rho g A$)
Loads

$-N(0)/qR$

	Exact (central)		Approx.	Exact (gravity)	
λ	$\alpha = 0.20$	$\alpha = 0.80$		$\alpha = 0.20$	$\alpha = 0.80$
1	0.6142	0.5974	0.6154	0.6069	0.4832
2	0.8646	0.8615	0.8649	0.8571	0.7410
4	0.9624	0.9618	0.9624	0.9548	0.8389
8	0.9903	0.9902	0.9903	0.9827	0.8666
10	0.9938	0.9937	0.9938	0.9862	0.8700

$10^2(M(0)/qL^2)$

	Exact (central)		Approx.	Exact (gravity)	
λ	$\alpha = 0.20$	$\alpha = 0.80$		$\alpha = 0.20$	$\alpha = 0.80$
1	4.807	4.769	4.808	4.814	4.845
2	1.686	1.641	1.689	1.697	1.791
4	0.4689	0.4530	0.4699	0.4803	0.6315
8	0.1206	0.1163	0.1209	0.1324	0.3028
10	0.0774	0.0746	0.0781	0.0893	0.2622

$10^2(EIw(0)/qL^4)$

	Exact (central)		Approx.	Exact (Gravity)	
λ	$\alpha = 0.20$	$\alpha = 0.80$		$\alpha = 0.20$	$\alpha = 0.80$
1	0.5022	0.5265	0.5008	0.5020	0.5266
2	0.1762	0.1802	0.1760	0.1762	0.1816
4	0.0490	0.0497	0.0490	0.0491	0.0523
8	0.0126	0.0127	0.0126	0.0128	0.0158
10	0.0081	0.0082	0.0081	0.0083	0.0114

major variables of interest given in Table 4.1 can be regarded as solid and widely applicable over broad, practical ranges of λ.

In Tables 4.4 and 4.5 we show several limiting cases that are derived from the analytical estimates of Table 4.1. It is evident from Table 4.4 that, for arches with very small rises for which $\lambda^2 \ll 1$, all of the quantities involved (i.e., the radial displacement, all three stress resultants, the axial stress, and the support reactions) tend to the results that would be expected for classic beam behavior. That is, as λ^2 becomes very small, an extremely shallow arch behaves exactly like a beam under a uniform line load q, even to the classic Mc/I form of the compressive stress at the arch crown.

TABLE 4.3

Exact and Approximate (Small α) Axial and Moment
Stress Resultants and Radial Displacements at the
Crown of *Clamped* Arches under Centrally Directed
and Gravity ($q = \rho g A$) Loads

	$-N(0)/qR$				
	Exact (central)		Approx.	Exact (gravity)	
λ	$\alpha = 0.20$	$\alpha = 0.80$		$\alpha = 0.20$	$\alpha = 0.80$
1	0.2065	0.1506	0.2105	0.2001	0.0544
2	0.5148	0.4978	0.5161	0.5086	0.4033
4	0.8100	0.8094	0.8101	0.8041	0.7165
8	0.9447	0.9453	0.9446	0.9389	0.8530
10	0.9639	0.9643	0.9639	0.9581	0.8722

	$10^2(M(0)/qL^2)$				
	Exact (central)		Approx.	Exact (gravity)	
λ	$\alpha = 0.20$	$\alpha = 0.80$		$\alpha = 0.20$	$\alpha = 0.80$
1	3.299	3.428	3.292	3.310	3.557
2	2.018	2.026	2.018	2.026	2.149
4	0.7899	0.7690	0.7918	0.7978	0.8853
8	0.2300	0.2209	0.2308	0.2374	0.3345
10	0.1502	0.1439	0.1507	0.1575	0.2572

	$10^2(EIw(0)/qL^4)$				
	Exact (central)		Approx.	Exact (gravity)	
λ	$\alpha = 0.20$	$\alpha = 0.80$		$\alpha = 0.20$	$\alpha = 0.80$
1	0.2069	0.2342	0.2053	0.2072	0.2461
2	0.1265	0.1350	0.1258	0.1267	0.1396
4	0.0495	0.0509	0.0494	0.0497	0.0533
8	0.0144	0.0146	0.0144	0.0145	0.0164
10	0.0094	0.0095	0.0094	0.0095	0.0112

Conversely, as we can see in Table 4.5, as λ^2 becomes very large, an arch behaves much as one would expect an arch or a ring segment of width b to respond when loaded by a pressure $p = q/b$ because then $qR/EA = (q/b)R/Eh$. In particular, the stress is uniformly compressive through the thickness, and the influence (or presence) of the moment, shear, and transverse boundary conditions is inconsequential. Further, the crown radial displacement takes on a form remarkably similar to that seen in thin rings and shells when the rise parameter takes on large values ($\lambda^2 \gg 1$), although it is interesting to note again that the centerline deflection of the clamped arch is larger than the corresponding deflection of the pinned arch. In fact, one measure of the stiffness of the loaded arch is the ratio of the pinned to the clamped crown

TABLE 4.4

Limiting Values of Crown Displacements, Stress Resultants, Stresses and Reactions for Large Rises (i.e., Small Values of the Arch Rise Parameter λ under Centrally Directed Loading)

	Pinned arches $(8\lambda^2/5 \ll 1)$	Clamped arches $(4\lambda^2/15 \ll 1)$
Radial displacement ($\theta = 0$)	$w^p(0) \cong \dfrac{5qL^4}{384EI}$	$w^c(0) \cong \dfrac{qL^4}{384EI}$
Axial resultant	$N^p(\theta) \cong 0$	$N^c(\theta) \cong 0$
Moment resultant ($\theta = 0$)	$M^p(\theta) \cong \dfrac{qL^2}{8}\left(1-\left(\dfrac{\theta}{\alpha}\right)^2\right)$	$M^c(\theta) \cong \dfrac{qL^2}{24}\left(1-3\left(\dfrac{\theta}{\alpha}\right)^2\right)$
Shear resultant ($\theta = 0$)	$Q^p(\theta) \cong -\dfrac{qL}{2}\left(\dfrac{\theta}{\alpha}\right)$	$Q^c(\theta) \cong -\dfrac{qL}{2}\left(\dfrac{\theta}{\alpha}\right)$
Axial stress ($z = h/2, \theta = 0$)	$\sigma^p_{crown} \cong -\underbrace{\left(\dfrac{qL^2}{8}\right)\left(\dfrac{h}{2}\right)\left(\dfrac{12}{bh^3}\right)}_{Mc/I}$	$\sigma^c_{crown} \cong -\underbrace{\left(\dfrac{qL^2}{24}\right)\left(\dfrac{h}{2}\right)\left(\dfrac{12}{bh^3}\right)}_{Mc/I}$
Horizontal reactions	$R^p_H = qR(8\lambda^2/5) \Rightarrow 0$	$R^c_H = qR(4\lambda^2/15) \Rightarrow 0$
Vertical reactions	$R^p_V = \dfrac{qL}{2}$	$R^c_V = \dfrac{qL}{2}$
Support moments	$M^p(\pm\alpha) = 0$	$M^c(\pm\alpha) = \mp\dfrac{qL^2}{12}$

displacement. That stiffness measure, derived from Equations (4.28) and (4.39), respectively, is

$$k_{arch} \sim \frac{w^p(0)}{w^c(0)} = 5\left(\frac{1+4\lambda^2/15}{1+8\lambda^2/5}\right) \qquad (4.70)$$

Equation (4.70) tells us that for very small values of λ, the crown deflection of the pinned arch/beam will be larger than for its clamped counterpart, indicating that the clamped arch beam is stiffer that the pinned beam, an unsurprising result. On the other hand, as λ gets large, the ratio in Equation (4.70) decreases, goes through the value 1 when $\lambda_{p=c} = \sqrt{15} \cong 3.873$, and then stabilizes in an asymptote:

$$k^{asymp}_{arch}\bigg|_{\lambda^2 \gg 1} = \frac{5}{6} \qquad (4.71)$$

TABLE 4.5

Limiting Values of Crown Displacements, Stress Resultants, Stresses and Reactions for Large Rises (i.e., Large Values of the Arch Rise Parameter λ under Centrally Directed Loading)

	Pinned arches $(8\lambda^2/5 \gg 1)$	Clamped arches $(4\lambda^2/15 \gg 1)$
Radial displacement ($\theta = 0$)	$w^p(0) = \left(\dfrac{25}{16}\right)\dfrac{qR^2}{EA}$	$w^c(0) = \left(\dfrac{15}{8}\right)\dfrac{qR^2}{EA}$
Axial resultant	$N^p(\theta) \cong -qR$	$N^c(\theta) \cong -qR$
Moment resultant ($\theta = 0$)	$M^p(\theta) \cong 0$	$M^c(\theta) \cong 0$
Shear resultant ($\theta = 0$)	$Q^p(\theta) \cong 0$	$Q^c(\theta) \cong 0$
Axial stress ($z = h/2$, $\theta = 0$)	$\sigma^p_{crown} \cong -\dfrac{qR}{A}$	$\sigma^c_{crown} \cong -\dfrac{qR}{A}$
Horizontal reactions	$R^p_H = qR$	$R^c_H = qR$
Vertical reactions	$R^p_V = qL/2$	$R^c_V = qL/2$
Support moments	$M^p(\pm\alpha) = 0$	$M^c(\pm\alpha) = -\dfrac{5qL^2}{16\lambda^2} \Rightarrow 0$

In addition to providing the correct (and thus reassuring) limiting cases, the estimates displayed in Table 4.1 also help us identify the transition from the bending behavior of curved beams to that of essentially uniform compression of "pure" arches. Clearly, that transition occurs as $8\lambda^2/5 \gg 1$ for pinned arches and as $4\lambda^2/15 \gg 1$ for clamped arches. There are several different judgments that are possible about just when that transition occurs.

For example, in parallel with the classic notion of an arch constituted of a set of masonry blocks that cannot sustain a tensile stress, the transition to arch behavior could be defined as occurring at the value of λ that ensures that the axial stress is in compression through the arch thickness. Thus, applying that condition to Equation (4.33) for pinned arches tells us that the transition point is

$$3\lambda^p_{trans} \le 8(\lambda^p_{trans})^2/5 \Rightarrow \lambda^p_{trans} = (\alpha^2 R/h)_{trans} \ge 1.88 \qquad (4.72)$$

Similarly, from Equation (4.43) for clamped arches, we see the transition occurring at

$$\lambda^c_{trans} \le 4(\lambda^c_{trans})^2/15 \Rightarrow \lambda^c_{trans} = (\alpha^2 R/h)_{trans} \ge 3.75 \qquad (4.73)$$

We can also express these transitions in terms of the arch rise itself. That is, in view of the definition (Equation 4.23) of the arch rise parameter, the transition of a pinned arch occurs at

$$3\lambda_{trans}^{p} \le 8(\lambda_{trans}^{p})^{2}/5 \Rightarrow f_{trans}^{p} \ge 0.94h \qquad (4.74)$$

and that of a clamped arch at

$$\lambda_{trans}^{c} \le 4(\lambda_{trans}^{c})^{2}/15 \Rightarrow f_{trans}^{c} \ge 1.88h \qquad (4.75)$$

Thus, we see that the transition to arch behavior occurs when the rise is essentially equal to the thickness!

We might also base our transition estimator on the idea that the dimensionless axial stress resultant might attain a preassigned fraction of a pure membrane state. For example, $|N(\theta)/qR| \cong 0.90$; that is, the axial stress resultant would be 90% of that membrane state. In this case, we would estimate the corresponding behavioral transition points for pinned and clamped arches respectively as

$$8(\lambda_{trans}^{p})^{2}/5 \cong 9 \Rightarrow \lambda_{trans}^{p} = (\alpha^{2}R/h)_{trans} \cong 2.37 \qquad (4.76)$$

and

$$4(\lambda_{trans}^{c})^{2}/15 \cong 9 \Rightarrow \lambda_{trans}^{c} = (\alpha^{2}R/h)_{trans} \cong 5.81 \qquad (4.77)$$

Thus, for clamped arches, the transition to an almost pure compressive state will occur for those arches with significantly larger rises.

Perhaps the most interesting results, especially in historical terms, are those we display in Table 4.6, wherein we list measured and calculated arch parameters for a dozen of Robert Maillart's deck-stiffened arch bridges. It is quite clear that the rise parameter λ calculated for each of these bridges is a very large number. In fact, and especially in the light of the immediately preceding discussion, there can be no doubt that the basic arches in Maillart's designs were very much in the uniform compressive state that he intended. It is also worth noting that the large values of λ also confirmed another of Maillart's intuitions: Namely, that the requisite membrane state in such an arch could be developed and sustained in relatively thin arches, which are more aesthetically pleasing and indicative of the force developed within them.

As noted before, Maillart analyzed his designs in a far simpler and more direct manner, and his intuitions were more than adequately confirmed. Interestingly enough, in Switzerland at that time, all designs were confirmed experimentally, not theoretically, as that was felt to be the most provident path to take for design in a public environment from the standpoints of the

TABLE 4.6

Values of Various Measured (Span, Rise, Thickness) and
Calculated (Radius, Semivertex Angle, Rise) Arch Parameters for
the Underlying Arches of Robert Maillart's Deck-Stiffened Arches

Bridge	L	f	h	R	h/R	α	λ
Flienglibach	38.7	5.17	0.25	38.8	0.0064	0.52	42.2
Schrähbach	28.8	4.02	0.18	27.8	0.0065	0.54	45.8
Valtschielbach	43.2	5.20	0.23	47.5	0.0048	0.47	46.1
Landquart	30.0	7.90	0.26	18.2	0.0143	0.97	65.8
Spital	30.0	3.26	0.24	36.1	0.0066	0.43	27.6
Landholz	26.0	3.40	0.16	26.6	0.0060	0.51	43.4
Hombach	21.0	3.00	0.16	19.9	0.0080	0.56	38.5
Luterstalden	12.5	2.55	0.16	8.9	0.0180	0.77	33.5
Traubach	40.0	5.60	0.20	38.5	0.0052	0.55	57.4
Bohlbach	14.4	2.70	0.16	11.0	0.0145	0.72	35.2
Schwandbach	37.4	6.00	0.20	32.1	0.0062	0.62	62.0
Töll	38.0	3.50	0.14	53.3	0.0026	0.36	50.6

governing public authorities and of the leading engineers of the time who recognized that their understanding of the issues was not yet established beyond question. Still further, and in this context rather ironically, Maillart reportedly believed "in the priority of design over analysis, in the subordinate role of analysis as a tool of design." In fact, in Maillart's own words, "Only a fully simplified method of analysis is…both possible and sufficient" (Billington 1979). Of course, it might be counterargued that Maillart's modeling ideas were confirmed only because his designs were very conservative in the sense that their values of λ were very large!

Our final observation concerns the three models used to describe arch behavior—that is, models based on, respectively, the curvature descriptions derived from the Sanders, Donnell, and shallow shell or arch theories. A comparison of all our results shows that the results produced by the Sanders and shallow models are, for small semivertex angles α, indistinguishable from one another. And while we did not show the details of the Donnell solutions, this situation obtained there as well. This strongly suggests that differences in arch behavior are far more sensitive to the formulation of the midsurface strain (Equation 4.4 evaluated at $z = 0$) than to changes in the curvature expression.

Shallow Curved Beams under End Loading

We have just shown that the linear response of arch structures to lateral (both gravitational and centrally directed) loading can be tracked via an arch rise parameter, λ, that is a function of the arch's semivertex angle α,

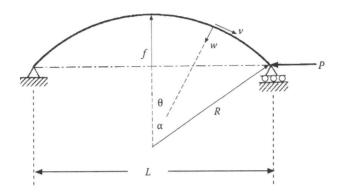

FIGURE 4.4
Geometry of and loading on a thin circular (pinned) curved beam (or arch-like) structure of radius R and semivertex angle α. (By permission of ASCE.)

thickness h, and radius R: When $\lambda^2 \ll 1$, the structure responds as a beam in bending, while for $\lambda^2 \gg 1$, the response is that of a classical arch in pure compression. It is interesting to ask how a similar arch-like structure would respond when subjected to a compressive axial—rather than lateral—load (Figure 4.4). Further, should such a structure be viewed as a curved beam or as a segment of a ring?

Clearly, the major difference is that one of the supports has to be allowed to move. In addition, while there is a geometric resemblance to a curved beam, the loading is not lateral, as we typically expect when we analyze beams. There is a sufficiently long history of work on curved beams and rings, and virtually all of it focuses on calculating deflections at discrete points of relatively thick rings by applying Castigliano's second theorem. We note that arches and curved beams are similar in both appearance (compare Figures 4.2 and 4.4) and underlying mathematical model of an axially compressed, curved, arch-like structure. Thus, as with the true arches we have just discussed, the axially loaded curved beam of Figure 4.4 can be analyzed with a linear model wherein the axial and bending responses are also coupled simply because of the curvature.

This stands in contrast to the analysis of straight beam-columns, wherein the coupling of axial and bending responses occurs due to a linearized axial load term in the transverse equilibrium equation. It also stands in contrast to the linearized and nonlinear stability analyses of thin arches. Further, we will show here that the behavior of such curved beams or arch-like structures under axial loading contrasts with that exhibited by the true arch: The curved beam responds as a straight axially loaded bar when $\lambda^2 \ll 1$ and as a beam in bending when $\lambda^2 \gg 1$.

So let us consider a planar curved beam of constant mean radius R and thickness h, loaded at one (moveable) end by a concentrated axial load P (Figure 4.4). The curved beam is taken as shallow, in which case the curvature

change may be approximated as in Equation (4.48). Then the first variation
of the total potential energy for such an arch-like structure is adapted from
(Equation 4.9) as

$$\delta^{(1)}\Pi = \int_{-\alpha}^{\alpha}\left\{\frac{N}{R}\left[\frac{d\delta v}{d\theta}-\delta w\right]-\frac{M}{R^2}\left(\frac{d^2\delta w}{d\theta^2}\right)\right\}R\,d\theta - P(\delta w(\alpha)\sin\alpha - \delta v(\alpha)\cos\alpha) \quad (4.78)$$

where the constitutive relations for the stress resultants in shallow arches are
written in terms of the displacements as

$$N(\theta) = \frac{EA}{R}\left(\frac{dv(\theta)}{d\theta}-w(\theta)\right)$$

$$M(\theta) = -\frac{EI}{R^2}\left(\frac{d^2w(\theta)}{d\theta^2}\right)$$

$$(4.79)$$

Note that Equation (4.78) reflects the constraint that the moveable support
translates only along the line of the horizontally applied load *P*—that is, we
allow no vertical movement. We find the equilibrium equations by setting
Equation (4.78) to zero so that, after the customary integration by parts, we
get the following:

$$\frac{dN(\theta)}{d\theta}=0$$

$$N(\theta)+\frac{1}{R}\frac{d^2M(\theta)}{d\theta^2}=0$$

$$(4.80)$$

This variational process also produces boundary conditions at the ends of
the arch, $\theta=\pm\alpha$. For the pinned arch considered here, four of the six required
conditions are familiar:

$$v(-\alpha)=0, \quad w(-\alpha)=0, \quad M(-\alpha)=0, \quad M(\alpha)=0 \quad (4.81)$$

Our fifth boundary condition states horizontal equilibrium at the loaded tip:

$$\frac{1}{R}\frac{dM(\alpha)}{d\theta}\sin\alpha - N(\alpha)\cos\alpha = P \quad (4.82)$$

while the sixth reflects constrained horizontal motion at the moveable support:

$$w(\alpha)+v(\alpha)\tan\alpha = 0 \quad (4.83)$$

We solve the equilibrium Equation (4.80) and use two moment boundary conditions (Equation 4.81) and the first boundary constraint (Equation 4.82) to find the normal stress resultant:

$$N(\theta) = -N_0 = -\frac{P}{\cos\alpha(1+\alpha\tan\alpha)} \tag{4.84}$$

and the moment resultant:

$$M(\theta) = (N_0 R/2)(\theta^2 - \alpha^2) \tag{4.85}$$

We find the corresponding tangential and radial displacements by integrating the corresponding constitutive Equation (4.79) under appropriate boundary (Equation 4.81) and constraint (Equation 4.83) conditions. Now, we find it convenient to introduce once again the arch rise parameter λ defined in Equation (4.23). With the aid of this definition, the tangential and radial displacements can be determined and written respectively as

$$v(\theta) = \left[\begin{array}{l} \lambda^2\left(-\dfrac{5}{2}\left(\dfrac{\theta}{\alpha}\right)+\left(\dfrac{\theta}{\alpha}\right)^3-\dfrac{1}{10}\left(\dfrac{\theta}{\alpha}\right)^5-\dfrac{8}{5}\right)-\left(\dfrac{\theta}{\alpha}+1\right) \\[2ex] +\dfrac{\alpha\tan\alpha}{2(1+\alpha\tan\alpha)}(1+8\lambda^2/5)\left(\dfrac{\theta}{\alpha}+1\right)^2 \end{array} \right]\left(\dfrac{N_0 R\alpha}{EA}\right) \tag{4.86}$$

and

$$w(\theta) = \left[\lambda^2\left(-\dfrac{5}{2}+3\left(\dfrac{\theta}{\alpha}\right)^2-\dfrac{1}{2}\left(\dfrac{\theta}{\alpha}\right)^4\right)+\left(\dfrac{\alpha\tan\alpha}{1+\alpha\tan\alpha}\right)(1+8\lambda^2/5)\left(\dfrac{\theta}{\alpha}+1\right)\right]\left(\dfrac{N_0 R}{EA}\right) \tag{4.87}$$

Interpreting Curved Beam Behavior

What can we learn from the preceding results about arch-like structure behavior, and how does it compare or contrast with the arch behavior we have already seen? First, as expected, the stress and displacement estimates are explicit functions of the arch rise parameter λ. Then we see that the axial stress resultant (Equation 4.84) becomes

$$N(\theta) = -N_0 \cong -\frac{P}{1+\alpha^2/2} \cong -P \tag{4.88}$$

We can use the result (Equation 4.88) and the definition of the rise parameter (Equation 4.23) to evaluate the moment resultant (Equation 4.85) at the curved beam's midpoint ($\theta = 0$), which then is

$$M(0) = -N_0 R \alpha^2 / 2 \cong -Pf \tag{4.89}$$

Equation (4.89) accurately reflects the moment produced by the (only) external load. We also see that the moment (Equation 4.89) vanishes as the arch rise tends to zero (i.e., as $f \to 0$).

The displacements at the midpoint of this curved beam are also interesting. From Equation (4.86) we see that, assuming small angles as usual, the tangential or in-plane displacement becomes

$$v(0) = -\left(\frac{2 + \alpha \tan \alpha}{2(1 + \alpha \tan \alpha)} \right)(1 + 8\lambda^2/5)\left(\frac{N_0 R \alpha}{EA} \right) \cong -(1 + 8\lambda^2/5)\left(\frac{PL}{2EA} \right) \tag{4.90}$$

Equation (4.90) clearly tells us that for end-loaded curved beams with very small rises, for which $\lambda^2 \ll 1$, the tangential displacement tends to the result that we would expect at the midpoint of an axially loaded straight bar; that is,

$$v(0)\big|_{\lambda^2 \ll 1} \cong -\left(\frac{PL}{2EA} \right) \tag{4.91}$$

This is quite unlike the laterally loaded arch results where we saw beamlike behavior in the small rise limit. The radial displacement at the midpoint (Equation 4.89) also exhibits similar behavior; that is,

$$w(0)\big|_{\lambda^2 \ll 1} \cong \left(\frac{\alpha \tan \alpha}{1 + \alpha \tan \alpha} \right)\left(\frac{N_0 R}{EA} \right) \cong \alpha^2 \frac{PR}{EA} \cong \alpha \frac{PL}{2EA} \tag{4.92}$$

We see in this small rise limit that the radial or out-of-plane displacement is smaller than the in-plane displacement by a factor of α, as we would expect for an axially loaded, nearly straight bar. It is also interesting here to calculate the axial stiffness of an almost straight, axially loaded bar, which we can define and calculate as

$$k \triangleq \frac{-P}{v(\alpha)/\cos\alpha} \equiv \frac{P}{w(\alpha)/\sin\alpha}$$

$$= \frac{EA}{L}\left(\frac{\cos^2\alpha(1 + \alpha \tan\alpha)^2}{1 + 8\lambda^2/\ 5} \right) \tag{4.93}$$

For small values of α, we then get

$$k \cong \frac{EA}{L}\left(\frac{1}{1+8\lambda^2/5}\right) \qquad (4.94)$$

Equation (4.94) obviously produces a familiar limit as $\lambda^2 \to 0$, and it also suggests that a slightly bent bar is less stiff than a straight bar. (We ought to bear in mind that the loaded tip is not free: It is constrained to move in a horizontal direction.) When $\lambda^2 \gg 1$, the axial stiffness gets much smaller, as would be expected for a (curved) beam for which bending—rather than extension—is the primary response.

Beyond that, the general behavior of the end-loaded bar as $\lambda^2 \gg 1$ is less clear. The midpoint circumferential displacement (Equation 4.90) becomes

$$v(0)\Big|_{\lambda^2 \gg 1} \cong -\frac{8\lambda^2}{5}\left(\frac{N_0 \alpha R}{EA}\right) \equiv -\frac{\alpha(Pf)L^2}{15EI} \qquad (4.95)$$

while the midpoint radial displacement (Equation 4.89) tends to

$$w(0)\Big|_{\lambda^2 \gg 1} \cong -\frac{5\lambda^2}{2}\left(\frac{N_0 R}{EA}\right) \equiv -\frac{5(Pf)L^2}{48EI} \qquad (4.96)$$

We can recognize in Equations (4.95) and (4.96) the bending response of a beam—note the flexibility coefficient of L^2/EI—to an applied moment proportional to Pf: A simple beam with a clockwise moment Pf at its right end produces a midpoint deflection of $w(0) = -PfL^2/16EI$, which is 60% of the value given by Equation (4.96).

Equation (4.95) also tells us that the tangential displacement at the midpoint is smaller than the transverse displacement by a factor of $(16\alpha/25)$, which would be consistent with expectations for such a bent beam. Further, Equations (4.92) and (4.96) show that the midpoint deflection changes sign and direction as λ^2 increases. Again, for small α, it can be shown that there is a transitional value of the rise parameter, $\lambda^2_{trans} \equiv 2\alpha^2/5$, for which the midpoint deflection vanishes (i.e., $w(0) = 0$), with the curved beam deflecting outward for $-\alpha \leq \theta < 0$ and inward for $0 < \theta \leq \alpha$. Similarly, the value of θ at which that transition occurs varies with both α and λ.

Finally, a transition point from bar-to-beam behavior can be identified by examining the circumferential stress of the arch; that is,

$$\sigma_{\theta\theta}(\theta, z) = \frac{N(\theta)}{A} + \frac{M(\theta)z}{I} \qquad (4.97)$$

If we evaluate Equation (4.97) at the curved beam's midpoint, using the results (Equations 4.84 and 4.85) and the definitions (Equations 4.23 and 4.24), we can write the axial normal stress (Equation 4.97) as

$$\sigma_{\theta\theta}(0,-h/2) \cong -\frac{P}{A}[1-3\lambda] \tag{4.98}$$

Thus, for $\lambda \ll 1/3$, the stress will be the purely compressive state expected in an arch, while for $\lambda \gg 1/3$, the stress changes sign as it reflects bending behavior. This is consistent with the results presented earlier.

Conclusions

In this chapter we explored the linear response of circular arch structures to centrally directed, "dead" pressure loading and to gravitational loading. We presented simple, closed-form estimates, useful for back-of-the-envelope analysis and design calculations; the major displacements; and stress resultants. We noted that essentially identical results are obtained when three different curvature expressions (e.g., Sanders, Donnell, and shallow) were used; the distinctions between the results were quite small, suggesting that the common midsurface or tangential strain expression is far more important than are differences in the curvature change expressions in the three kinematic models.

Further, these design-and-analysis estimates were shown to be as valid for gravitational loading as they are for centrally directed loads for shallow arches.

Using these estimates, we demonstrated that arch behavior ranges from elementary beam bending, for very shallow arch structures, to a pure compressive state for steep arches. The change in response can be tracked with an arch rise parameter that is a function of the arch's semivertex angle, thickness, and, radius: $\lambda = \alpha^2 R/h$. The transition from beam to arch behavior occurs for $\lambda^p_{trans} \sim 2$ for pinned arches and for $\lambda^c_{trans} \sim 4$ for clamped arches. Our theoretical analysis also showed that Robert Maillart's inspired design concept performed exactly as he intended in that the compressive states he posited were clearly and completely created in elegant slender arches.

Finally, we saw that the linear response of shallow curved beams (or arch-like structures) to an axial end load ranges from compression of a straight bar, for very shallow arch-like structures, to a bending state for steep curved beams. This behavior stands in stark contrast to that seen in laterally loaded true arches when tracked with the same geometric rise parameter λ. Further, since our estimates of the structures' behavior explicitly showed how the axial and bending stresses (and the displacements) depend on that rise parameter, we could readily identify the transition from extensional to bending response in that end-loaded curved beam.

Bibliography

Armenàkas, A. E. 2006. *Advanced mechanics of materials and applied elasticity.* Boca Raton, FL: CRC/Taylor & Francis.

Billington, D. P. 1979. *Robert Maillart's bridges: The art of engineering.* Princeton, NJ: Princeton University Press.

Boresi, A. P., and R. J. Schmidt. 2003. *Advanced mechanics of materials,* 6th ed. New York: John Wiley & Sons.

Budynas, R. G. 1999. *Advanced strength and applied stress analysis,* 2nd ed. New York: McGraw-Hill.

Dym, C. L. 1990. *Introduction to the theory of shells.* New York: Hemisphere Publishing.

———. 2002. *Stability theory and its applications to structural mechanics.* Mineola, NY: Dover Publications.

Dym, C. L., and I. H. Shames. 1973. *Solid mechanics: A variational approach.* New York: McGraw-Hill.

Dym, C. L., and H. E. Williams. 2011. Stress and displacement estimates for arches. *Journal of Structural Engineering* 137 (1): 48–59.

Heyman, J. 1995. *The stone skeleton: Structural engineering of masonry architecture.* Cambridge, England: Cambridge University Press.

Hoff, N. J. 1956. *The analysis of structures.* New York: John Wiley & Sons.

Sanders, J. L. 1963. Nonlinear theories for thin shells. *Quarterly of Applied Mathematics* 21: 21.

Schreyer, H. L., and E. F. Masur. 1966. Buckling of shallow arches. *Journal of the Engineering Mechanics Division,* ASCE 92 (EM4): 1–19.

Problems

4.1 Show that Equation (4.83) properly reflects the geometric constraint provided by the roller at the right support of the shallow curved beam of Figure 4.4.

4.2 Carry out the unshown variational details of Equation (4.78) to show that Equations (4.81)–(4.83) are the proper boundary conditions for the shallow curved beam shown in Figure 4.4.

4.3 Introducing the shear stress resultant $Q(\theta)$ defined in Equation (4.11), shows that requiring it to be directed radially inward on a positive transverse area of the curved beam (i.e., its normal being in the positive θ-direction) is consistent with Equation (4.12).

4.4 Show that a counterclockwise moment $M(\theta)$ on a positive transverse area of the curved beam makes Equation (4.11) a curved beam's moment equilibrium equation. (*Hint:* Note the radial boundary condition of Problem 4.2 becomes $Q(\alpha) = P\sin \alpha$, while the differential Equation (4.80) becomes $Q'(\alpha) + N(\alpha) = 0$, which is consistent with Equation (4.12) with $q_r = 0$.)

4.5 Consider a full circular ring ($0 \le \theta \le 2\pi$) acted on by n inwardly directed radial loads P that produce a net inward radial (only) displacement

δ_r. If the number of such loads is quite large (i.e., $n \gg 1$), the ring can be modeled as a set of n shallow curved beam segments, so Equation (4.80) is applicable. For a typical beam segment in which ($-\alpha \le \theta \le \alpha$), determine the radial and tangential displacements, respectively, $w(\theta)$ and $v(\theta)$, that satisfy $w(\alpha) = \delta_r$, $v(\alpha) = 0$, and $w'(\alpha) = 0$. (*Hint:* Are $w(\theta)$ and $v(\theta)$ odd or even in θ?)

4.6   Determine a relation between the load P and the radial displacement δ_r of Problem 4.5 by using symmetry and periodicity applied to its beam segment and thus show that force equilibrium requires that $P = -2N(\theta = 0)\sin \alpha \cong 2N_0\alpha$. What does symmetry require of $Q(\theta = 0)$?

4.7   Given the results of Problems 4.5 and 4.6, show that for a large ($n \gg 1$) number of shallow beam segments, the individual Ps can be expressed in terms of an equivalent force per unit length of circumference, $P = p \times 2R\alpha$, and that we can recover the equation for the inward deflection of a ring subject to a uniform line load:
$$\delta_r = \lim_{\substack{n \to \infty \\ \alpha \to 0}} \delta_r(P,\alpha) = pR^2 / EA.$$

4.8   Find the radial displacement under a radial concentrated force P applied at the apex of a shallow curved beam that is pinned at both ends (i.e., $\theta = \pm\alpha$).

4.9   Find the radial displacement under a radial concentrated force P applied at the apex of a shallow curved beam that is clamped at both ends (i.e., $\theta = \pm\alpha$).

4.10 Consider a shallow curved beam such as that depicted in Figure 4.4, except that both ends are free to move inward horizontally (on rollers), but unable to move vertically. Find the radial and tangential displacements for that curved beam when it is loaded by inwardly directed loads P as well as pinned at both ends (i.e., $\theta = \pm\alpha$).

4.11 Find the radial and tangential displacements of the shallow curved beam described in Problem 4.10 when it is subjected to positive (as defined by the sign convention of Problem 4.4) moments M_0 at both ends (i.e., $\theta = \pm\alpha$).

4.12 Show how the results of Problems 4.10 and 4.11 can be used to derive the influence coefficients for a shallow curved beam that reflect Maxwell's reciprocity principle.

5

Estimating the Behavior of
Coupled Discrete Systems

Summary

In this chapter we will talk about modeling the behavior of coupled discrete linear systems, first introducing an extension of the famous Castigliano's theorems to show that we can (1) formally apportion the loads among the elements of a coupled *parallel* system by minimizing that parallel system's displacement as determined by Castigliano's second theorem, and (2) formally apportion the displacement among the elements of a coupled *series* system by minimizing that series system's load as determined from Castigliano's first theorem. These extensions provide a means for apportioning loads in coupled continuous systems, which we will show in Chapter 6 for two systems of coupled Timoshenko beams supporting discrete and continuous loads. We then conclude this chapter by describing a second classic mechanics result, the Rayleigh quotient, and using it to lay the foundation for estimating the natural frequencies of elastic elements, which we will do in Chapter 7.

Introduction

Ultimately, engineering problems must be rendered discrete to be solved—meaning that we have to express even our most elegant analytical results in terms of discrete numbers or values if we want to solve real problems. Almost every approximate technique and all numerical methods—whether finite difference or finite element—wind up in a discrete, rather than continuous, formulation. Furthermore, we often hone our intuition and learn a lot by formulating and solving discrete problems. Structural mechanics is classic in this sense because we often talk about the stiffness of structural elements as if they were springs with sometimes complicated spring constants that reflect various aspects of the geometry of loading. Think, for example, of the tip deflection of a tip-loaded cantilever as being represented as the extension of a spring of stiffness $k_{cant} = 3EI/L^3$. We will say more about this in Chapter 6, but suffice it for now to say that we can model buildings as collections of beam components for both preliminary design and the estimation of natural frequencies.

In static models we typically assume *series* models of behavior wherein bending and shear deflections of beams add to form the total deflection. However, while we add the shear and bending deflections of individual beam elements *series,* we assemble the beam components in *parallel* for a building because we require that the building deflection be the same for each of its component elements, whether those elements are individually in (pure) shear or bending or a series combination of the two. We model the dynamic response of a building composed of a shear core and a bent tube, for example, by a parallel coupling of an Euler beam to a companion shear beam. In these models, since we constrain the component deflections to equal one another (and to the total deflection), the bending and shear stiffnesses add linearly, and the external load is distributed among the individual components in proportion to each component's stiffness as a fraction of the total stiffness.

We can easily apportion an external load among discrete parallel components where there is a single "concentrated" external load, but it is not so clear how we allocate a load that is spatially distributed over a beam's length (or a building model's height). Here, by analogy, we turn to the discretization inherent in Castigliano's second theorem in which we can easily calculate the deflection under a concentrated load, and we similarly find the deflection under a distributed load by introducing a phantom force. So we will describe an extension of Castigliano's second theorem that allows us to calculate load apportionment and consequent deflections for discrete models. We will do similar calculations for continuous models in Chapter 6, where the individual or component beams are Timoshenko beams in which both bending and shear deflections are added in series, which we then aggregate in parallel constructions to simulate whole buildings.

We will close this (short!) chapter by describing another classic of the applied mechanics—namely, the Rayleigh quotient. We will show it to be a simple, powerful estimator of the natural frequencies of rather complicated systems. Part of its power resides in the fact that we can get very good estimates of these frequencies with rather limited guesses or information about the systems we will be modeling. Then, in Chapter 7, we will show how the Rayleigh quotient can be used to get some really nice estimates of the dynamics of building models.

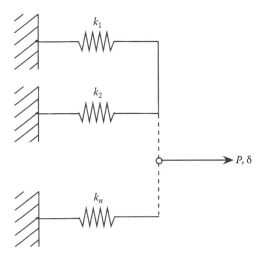

FIGURE 5.1
A set of springs arrayed in *parallel*, thus having a common endpoint displacement δ.

Extending the Castigliano Theorems to Discrete Systems

We start with an elementary discrete system of two linear springs ostensibly connected in parallel (Figure 5.1). If each spring has stiffness k_i and supports an internal spring force F_i, the complementary energy contained in the system is

$$U^* = \sum_{i=1}^{2} \frac{F_i^2}{2k_i} \tag{5.1}$$

As we apply an external load P to the system, we assume that the respective internal spring forces are

$$F_1 \equiv rP \quad \text{and} \quad F_2 = (1-r)P \tag{5.2}$$

where r is a positive constant such that $0 \le r \le 1$. Then we can write the system's complementary energy (Equation 5.1) as

$$U^* = \frac{1}{2}\left(\frac{r^2}{k_1} + \frac{(1-r)^2}{k_2}\right)P^2 \tag{5.3}$$

We then apply Castigliano's second theorem by minimizing the total complementary energy Π^* to find the resulting compatible system displacement:

$$\delta_{par} = \frac{dU^*}{dP} = \left(\frac{r^2}{k_1} + \frac{(1-r)^2}{k_2} \right) P \tag{5.4}$$

What is r and what is its value? In fact, r is the *apportionment factor* whose value we find by minimizing the system displacement δ_{par}—that is, the value of r for which

$$\frac{d\delta_{par}}{dr} = 2\left(\frac{r}{k_1} - \frac{(1-r)}{k_2} \right) P = 0 \tag{5.5}$$

which is

$$r = \frac{k_1}{k_1 + k_2} \tag{5.6}$$

and from which we see that

$$\delta_{par} = \frac{P}{k_1 + k_2} \tag{5.7}$$

We recognize Equation (5.7) as a familiar and expected result, although the way in which we have derived it is not. Further, we also see that the value of r in Equation (5.5) creates a minimum in δ_{par} since

$$\frac{d^2\delta_{par}}{dr^2} = 2\left(\frac{1}{k_1} + \frac{1}{k_2} \right) P > 0 \tag{5.8}$$

Thus, we postulate the following extension of Castigliano's second theorem:

> The minimum of the (compatible) displacement of a parallel system as determined by Castigliano's second theorem corresponds to an equilibrium apportioning of the loads carried by the elements in that system.

We interpret this result in physical terms as follows. Note that the system displacement (Equation 5.7) found by this process means that each spring has the same displacement; that is,

$$\delta_1 = \left(\frac{F_1 = rP}{k_1} \right) \equiv \delta_{par} \quad \text{and} \quad \delta_2 = \frac{(F_2 = (1-r)P)}{k_2} \equiv \delta_{par} \tag{5.9}$$

The value of r thus determined also guarantees that the displacement of each element in this parallel system will be the same. Thus, it ought to be no surprise that the apportionment factor thus determined produces a familiar, seemingly self-evident result.

We can readily extend this extension of Castigliano's second theorem to systems with n degrees of freedom, in which case we write the corresponding complementary energy in terms of n distribution factors for which $\sum_{i=1}^{n} r_i = 1$,

$$U_n^* = \sum_{i=1}^{n} \frac{F_i^2}{2k_i} = \frac{1}{2}\left(\frac{r_1^2}{k_1} + \frac{r_2^2}{k_2} + \cdots + \frac{(1 - r_1 - r_2 \cdots - r_{n-1})^2}{k_n} \right) P^2 \qquad (5.10a)$$

which we can also write in the form of

$$U_n^* = \left(\frac{k_n}{k_1} r_1^2 + \frac{k_n}{k_2} r_2^2 + \cdots + (r_{n-1} + r_{n-2} + \cdots + r_1 - 1)^2 \right) \frac{P^2}{2k_n} \qquad (5.10b)$$

Once again, Castigliano's second theorem produces the system displacement, in this case:

$$\delta_{par/n} = \frac{dU_n^*}{dP} = \left(\frac{k_n}{k_1} r_1^2 + \frac{k_n}{k_2} r_2^2 + \cdots + (r_n + r_{n-1} + \cdots + r_1 - 1)^2 \right) \frac{P}{k_n} \qquad (5.11)$$

We minimize the system displacement $\delta_{par/n}$ with respect to the factors $r_1, r_2, \ldots r_{n-1}$ to produce the following system of n equations for those factors:

$$
\begin{bmatrix}
(k_n/k_1 + 1) & 1 & \cdots & 1 & 1 \\
1 & (k_n/k_2 + 1) & \cdots & 1 & 1 \\
1 & 1 & \ddots & 1 & 1 \\
1 & 1 & \cdots & (k_n/k_{n-2} + 1) & 1 \\
1 & 1 & \cdots & 1 & (k_n/k_{n-1} + 1)
\end{bmatrix}
\begin{Bmatrix}
r_1 \\
r_2 \\
\vdots \\
r_{n-2} \\
r_{n-1}
\end{Bmatrix}
=
\begin{Bmatrix}
1 \\
1 \\
\vdots \\
1 \\
1
\end{Bmatrix}
$$
$$(5.12)$$

We see by inspection that the (unique) solution to this linear system is

$$
\begin{Bmatrix}
r_1 \\
r_2 \\
\vdots \\
r_{n-2} \\
r_{n-1}
\end{Bmatrix}
=
\frac{1}{\sum_{i=1}^{n} k_i}
\begin{Bmatrix}
k_1 \\
k_2 \\
\vdots \\
k_{n-2} \\
k_{n-1}
\end{Bmatrix}
\qquad (5.13)
$$

because when we substitute Equation (5.13) into Equation (5.12), we get

$$
\begin{bmatrix}
(k_n/k_1 + 1) & 1 & \cdots & 1 & 1 \\
1 & (k_n/k_2 + 1) & \cdots & 1 & 1 \\
1 & 1 & \ddots & 1 & 1 \\
1 & 1 & \cdots & (k_n/k_{n-2} + 1) & 1 \\
1 & 1 & \cdots & 1 & (k_n/k_{n-1} + 1)
\end{bmatrix}
\left(\frac{1}{\displaystyle\sum_{i=1}^{n} k_i} \right)
\left\{
\begin{matrix}
k_1 \\ k_2 \\ \vdots \\ k_{n-2} \\ k_{n-1}
\end{matrix}
\right\}
$$

(5.14)

$$
= \left(\frac{1}{\displaystyle\sum_{i=1}^{n} k_i} \right)
\left\{
\begin{matrix}
(k_n + k_1) + k_2 + \cdots + k_{n-2} + k_{n-1} \\
k_1 + (k_n + k_2) + \cdots + k_{n-2} + k_{n-1} \\
k_1 + k_2 + \cdots + k_n + \cdots + k_{n-2} + k_{n-1} \\
k_1 + k_2 + \cdots + (k_n + k_{n-2}) + k_{n-1} \\
k_1 + k_2 + \cdots + k_{n-2} + (k_n + k_{n-1})
\end{matrix}
\right\}
\equiv
\left\{
\begin{matrix}
1 \\ 1 \\ \vdots \\ 1 \\ 1
\end{matrix}
\right\}
$$

Equation (5.13) clearly shows that the load apportionment factors for each of the n elements in the system are, in scalar form,

$$
r_j = \frac{k_j}{\displaystyle\sum_{i=1}^{n} k_i} \qquad j = 1, 2 \cdots n - 1
$$

(5.15)

Finally, we note that the Hessian matrix of the quadratic form (Equation 5.11) that makes up the displacement $\delta_{par/n}$ is clearly a positive constant; that is,

$$
\begin{bmatrix}
(k_n/k_1 + 1) & 1 & \cdots & 1 & 1 \\
1 & (k_n/k_2 + 1) & \cdots & 1 & 1 \\
1 & 1 & \ddots & 1 & 1 \\
1 & 1 & \cdots & (k_n/k_{n-2} + 1) & 1 \\
1 & 1 & \cdots & 1 & (k_n/k_{n-1} + 1)
\end{bmatrix}
> 0 \quad (5.16)
$$

Hence, the displacements of each element are the same minimum displacement:

$$
\delta_i = \left(\frac{F_i = r_i P}{k_i} \right) \equiv \delta_{par/n}
$$

(5.17)

FIGURE 5.2
A set of springs arrayed in *series*, thus supporting a common applied force P.

We note again that, in contrast with traditional descriptions in which a common displacement for parallel systems is imposed ab initio as a constraint, this extension of Castigliano's second theorem has that common displacement emerging as an end result.

We can also develop a corresponding extension of Castigliano's first theorem for the displacements that result when a common load is applied to a system of springs in series (Figure 5.2). Thus, if each (of two, for now) spring has stiffness k_i and extends an amount δ_i, the strain energy stored in the system is

$$U = \sum_{i=1}^{2} \frac{1}{2} k_i \delta_i^2 \qquad (5.18)$$

If we apply an external load P to the system, we can assume the respective spring displacements to be

$$\delta_1 \equiv r\delta \quad \text{and} \quad \delta_2 = (1-r)\delta \qquad (5.19)$$

where r is again a positive constant such that $0 \le r \le 1$. Then the system's stored energy (Equation 5.1) becomes

$$U = \frac{1}{2}\left(k_1 r^2 + k_2(1-r)^2\right)\delta^2 \qquad (5.20)$$

Castigliano's first theorem states that the resulting force supported by the system is

$$P_{ser} = \frac{dU}{d\delta} = (k_1 r^2 + k_2(1-r)^2)\delta \qquad (5.21)$$

As with our preceding discussion (following Equation 5.4), it is well known that Equation (5.21) represents the statement of Castigliano's first theorem, which can be derived by minimizing the total potential energy Π. Thus, the displacement that minimizes the total potential energy produces the equilibrium statement as given by Equation (5.21).

Also similarly, we determine the value of the apportionment factor r as that which minimizes the system load P_{ser}, that is, the one for which

$$\frac{dP_{ser}}{dr} = 2(k_1 r - k_2(1-r))\delta = 0 \tag{5.22}$$

which results in

$$r = \frac{k_2}{k_1 + k_2} \tag{5.23}$$

and from which it follows that

$$P_{ser} = \left(\frac{1}{k_1} + \frac{1}{k_2}\right)^{-1}\delta \tag{5.24}$$

Again, Equation (5.24) is familiar and expected, and here too we see that the value of r in Equation (5.23) actually corresponds to a minimum of P_{ser} since

$$\frac{d^2 P_{ser}}{dr^2} = 2(k_1 + k_2)\delta > 0 \tag{5.25}$$

Thus, an extension of Castigliano's first theorem now emerges:

> The minimum of the (equilibrium) force in a series system as determined by Castigliano's first theorem corresponds to a compatible apportioning of the displacements of the elements in that system.

A physical interpretation of this result is also possible. Note that the system load (Equation 5.24) found by this process means that each spring supports or transmits the same load; that is,

$$F_1 = k_1(\delta_1 = r\delta) = P_{ser} \quad \text{and} \quad F_2 = k_2(\delta_2 = (1-r)\delta) = P_{ser} \tag{5.26}$$

The value of r we have just determined also guarantees that the force carried by each element in this parallel system will be the same. Thus, we should not be surprised that the apportionment factor thus found yields still another familiar, seemingly self-evident result.

Formally Proving the Castigliano Theorem Extensions

We now present a formal proof of the solution (Equation 5.13) as an interesting application of matrix arithmetic. We start by writing the basic Equation (5.12) in matrix form as

$$\mathbf{Kr} = \mathbf{v} \qquad (5.27)$$

Here $\mathbf{r}$ and $\mathbf{v}$ are column matrices with $n - 1$ rows,

$$\mathbf{r} \triangleq [r_1 \quad r_2 \quad \cdots \quad r_{n-1} \quad r_{n-2}]^T, \quad \mathbf{v} \triangleq [1 \quad 1 \quad \cdots \quad 1 \quad 1]^T \qquad (5.28)$$

and we define a symmetric stiffness matrix $\mathbf{K}$ of order $n - 1$ as

$$\mathbf{K} \triangleq \begin{bmatrix} (k_n/k_1 + 1) & 1 & \cdots & 1 & 1 \\ 1 & (k_n/k_2 + 1) & \cdots & 1 & 1 \\ 1 & 1 & \ddots & 1 & 1 \\ 1 & 1 & \cdots & (k_n/k_{n-2} + 1) & 1 \\ 1 & 1 & \cdots & 1 & (k_n/k_{n-1} + 1) \end{bmatrix} \qquad (5.29)$$

Note that we can subtract the quantity 1 from every element of the stiffness matrix (Equation 5.29), so we can rewrite it as the following sum:

$$\mathbf{K} = \mathbf{D} + \mathbf{vv}^T \qquad (5.30)$$

where $\mathbf{D}$ is a diagonal matrix of order $n - 1$,

$$\mathbf{D} \triangleq \begin{bmatrix} k_n/k_1 & 0 & \cdots & 0 & 0 \\ 0 & k_n/k_2 & \cdots & 0 & 0 \\ 0 & 0 & \ddots & 0 & 0 \\ 0 & 0 & \cdots & k_n/k_{n-2} & 0 \\ 0 & 0 & \cdots & 0 & k_n/k_{n-1} \end{bmatrix} \qquad (5.31)$$

In view of Equations (5.28) and (5.31), we can write Equation (5.29) as

$$\mathbf{K} = \mathbf{D}(\mathbf{I} + \mathbf{D}^{-1}\mathbf{vv}^T) \triangleq \mathbf{D}(\mathbf{I} + \mathbf{uv}^T) \qquad (5.32)$$

where we have introduced $\mathbf{I}$ as the unit matrix of order $n - 1$, as well as a new variable, $\mathbf{u}$,

$$\mathbf{u} \triangleq \mathbf{D}^{-1}\mathbf{v} = (1/k_n)\mathbf{k} \qquad (5.33)$$

and where in parallel with Equation (5.28), we now see that $\mathbf{k}$ is a column matrix with $n - 1$ rows:

$$\mathbf{k} \triangleq [k_1 \quad k_2 \quad \cdots \quad k_{n-1} \quad k_{n-2}]^T \qquad (5.34)$$

Now, the matrix $\mathbf{D}$ in Equation (5.34) is invertible since $k_i > 0$ for $i = 1, 2 \ldots n$. Further, since

$$\mathbf{u}\mathbf{v}^T = \mathbf{D}^{-1}\mathbf{v}\mathbf{v}^T = \frac{1}{k_n}\begin{bmatrix} k_1 & k_1 & \cdots & k_1 & k_1 \\ k_2 & k_2 & \cdots & k_2 & k_2 \\ \cdots & \cdots & \cdots & \cdots & \cdots \\ k_{n-2} & k_{n-2} & \cdots & k_{n-2} & k_{n-2} \\ k_{n-1} & k_{n-1} & \cdots & k_{n-1} & k_{n-1} \end{bmatrix} \qquad (5.35)$$

it can be shown that

$$\det(\mathbf{I} + \mathbf{u}\mathbf{v}^T) = \det(1 + \mathbf{v}^T\mathbf{u}) = 1 + \mathbf{v}^T\mathbf{u} \qquad (5.36)$$

where $\mathbf{v}^T\mathbf{u}$ is a *scalar* that is readily evaluated from Equations (5.28) and (5.33); that is,

$$\mathbf{v}^T\mathbf{u} = k_n^{-1}\sum_{i=1}^{n-1}k_i \equiv k_n^{-1}\sum_{i=1}^{n}k_i - 1 \qquad (5.37)$$

so that the matrix $(\mathbf{I} + \mathbf{u}\mathbf{v}^T)$ is also invertible. Then, by direct calculation, we see that

$$(\mathbf{I} + \mathbf{u}\mathbf{v}^T)^{-1} = \left(\mathbf{I} - \frac{\mathbf{u}\mathbf{v}^T}{1+\mathbf{v}^T\mathbf{u}}\right) = \left(\frac{(1+\mathbf{v}^T\mathbf{u})\mathbf{I} - \mathbf{u}\mathbf{v}^T}{1+\mathbf{v}^T\mathbf{u}}\right) = \left(\frac{\left(k_n^{-1}\sum_{i=1}^{n}k_i\right)\mathbf{I} - \mathbf{u}\mathbf{v}^T}{k_n^{-1}\sum_{i=1}^{n}k_i}\right) \qquad (5.38)$$

We then find the inverse of the matrix (Equation 5.33) by substituting Equation (5.37) therein and performing the required inversion; that is,

$$\mathbf{K}^{-1} = \left(\mathbf{I} + \mathbf{uv}^T\right)^{-1}\mathbf{D}^{-1} = \left(\frac{\left(k_n^{-1}\sum\limits_{i=1}^{n}k_i\right)\mathbf{I} - \mathbf{uv}^T}{k_n^{-1}\sum\limits_{i=1}^{n}k_i}\right)\mathbf{D}^{-1} = \mathbf{D}^{-1} - \frac{k_n\mathbf{uv}^T}{n}\mathbf{D}^{-1} \quad (5.39)$$

where the inverse of the diagonal matrix $\mathbf{D}$ is

$$\mathbf{D}^{-1} = \frac{1}{k_n}\begin{bmatrix} k_1 & 0 & \cdots & 0 & 0 \\ 0 & k_2 & \cdots & 0 & 0 \\ 0 & 0 & \ddots & 0 & 0 \\ 0 & 0 & \cdots & k_{n-2} & 0 \\ 0 & 0 & \cdots & 0 & k_{n-1} \end{bmatrix} \quad (5.40)$$

Then the load apportionment factors are

$$\mathbf{r} = \left[\mathbf{D}^{-1} - \frac{k_n\mathbf{uv}^T}{\sum\limits_{i=1}^{n}k_i}\mathbf{D}^{-1}\right]\mathbf{v} \quad (5.41)$$

We can then use Equations (5.36) and (5.41) to show that

$$\frac{k_n\mathbf{uv}^T}{\sum\limits_{i=1}^{n}k_i}\mathbf{D}^{-1}\mathbf{v} = \frac{\sum\limits_{i=1}^{n-1}k_i}{k_n\sum\limits_{i=1}^{n}k_i}\mathbf{k} = \frac{\sum\limits_{i=1}^{n}k_i - k_n}{k_n\sum\limits_{i=1}^{n}k_i}\mathbf{k} \quad (5.42)$$

And lastly, after we substitute Equations (5.40) and (5.41) into Equation (5.42), we find that the load apportionment factors are exactly those in Equation (5.13):

$$\mathbf{r} = \left(1 / \sum\limits_{i=1}^{n}k_i\right)\mathbf{k} \quad (5.43)$$

Clearly, the result (Equation 5.43) demonstrates that we have formally proved our extension of Castigliano's second theorem!

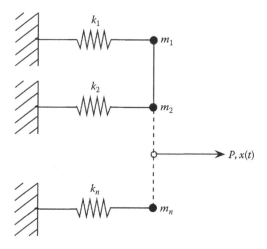

FIGURE 5.3
A set of springs and masses arrayed in *parallel,* thus having a common endpoint motion *x(t).*

Rayleigh Quotients for Discrete Systems

We now describe one of the more renowned techniques or principles of vibration theory. Consider a simple system of a single mass attached to the free end of a single spring (see either Figure 5.3 or 5.4 with $n = 1$). It is easy to write out the kinetic and stored energies for the motion $x(t)$ of the mass m at the end of the spring whose stiffness is k:

$$T_1 = \frac{1}{2}m\dot{x}^2(t) \quad \text{and} \quad V_1 = \frac{1}{2}kx^2(t) \tag{5.44}$$

Lord J. W. S. Rayleigh made the observation that when such an ideal system is vibrating freely—that is, with no damping and without any driving force—the energy in the system would oscillate between maximum kinetic energy and, half a cycle later, maximum kinetic energy. Modeling

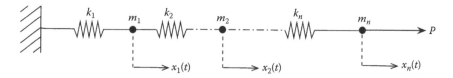

FIGURE 5.4
A set of springs and masses arrayed in *series.*

that mathematically means that if we assume the free motion is described by

$$x(t) = A \cos \omega_1 t \tag{5.45}$$

the maximum energies are

$$T_{1max} = \frac{1}{2} m A^2 \omega_1^2 \quad \text{and} \quad V_{1max} = \frac{1}{2} k A^2 \tag{5.46}$$

and the equation of the two energy maxima provides a formula for the natural or fundamental frequency at which the system is vibrating:

$$\omega_1^2 = \frac{k A^2}{m A^2} = \frac{k}{m} \tag{5.47}$$

We observe a couple of things in this remarkably simple—and almost universally known—formula. First, this final result says that natural frequency is a function—actually a ratio or a quotient—only of the system's stiffness and its mass. Second (and this was Rayleigh's insightful recognition), the frequency quotient is independent of the vibration amplitude! Thus, we term Equation (5.47) as the *Rayleigh quotient*, albeit the simplest possible such quotient.

We will now examine what happens to the Rayleigh quotient for more complex discrete systems in order to lay a brief foundation for applying the quotient, also called *Rayleigh's principle*, to continuous systems, which we will do in Chapter 7. So suppose we were modeling some structural system as a parallel system of spring-mass oscillators, as depicted in Figure 5.3. Because of the constraint that all the systems have the same net endpoint deflection, we have only one degree of freedom. We illustrate what this means by writing the counterparts of Equation (5.45) for n such oscillators in parallel:

$$T_{par} = \frac{1}{2} \left(\sum_{i=1}^{n} m_i \right) \dot{x}^2(t) \quad \text{and} \quad V_{par} = \frac{1}{2} \left(\sum_{i=1}^{n} k_i \right) x^2(t) \tag{5.48}$$

We can then easily find the Rayleigh quotient for a parallel system that corresponds to Equation (5.48) as

$$\omega_{par}^2 = \frac{\displaystyle\sum_{i=1}^{n} k_i}{\displaystyle\sum_{i=1}^{n} m_i} \tag{5.49}$$

We should note that the stiffness of the coupled system is simply the sum of the stiffnesses of each component, and the mass of the coupled system is similarly the sum of the component masses. We also see that Equation (5.49) predicts only one value for the frequency, ω^2_{par}, which is a direct consequence of the system having only one degree of freedom. Equation (5.49) seems such an obvious result—after all, for a parallel system we just add the masses and the stiffnesses!—that it is natural to wonder why the fuss is made over the Rayleigh quotient.

We justify the fuss when we consider *series* couplings of spring-mass oscillators, as in Figure 5.4. Starting again with the simple case of n oscillators, we note that there are n degrees of freedom and that the kinetic and potential energies now look like

$$T_{ser} = \frac{1}{2}\sum_{i=1}^{n} m_i \dot{x}_i^2(t) \quad \text{and} \quad V_{ser} = \frac{1}{2}\sum_{i=1}^{n} k_i \left(x_i(t) - x_{i-1}(t) \right)^2 \tag{5.50}$$

If we introduce displacement, mass, and stiffness matrices as, respectively,

$$\mathbf{X} = [x_1(t) \quad x_2(t) \quad \cdots \quad x_n(t)]^T \tag{5.51a}$$

$$\mathbf{M} = \begin{bmatrix} m_1 & 0 & 0 & 0 \\ 0 & m_2 & 0 & 0 \\ 0 & 0 & \ddots & 0 \\ 0 & 0 & 0 & m_n \end{bmatrix} \tag{5.51b}$$

and

$$\mathbf{K} = \begin{bmatrix} k_1 + k_2 & -k_2 & 0 & 0 \\ -k_2 & k_2 + k_3 & -k_3 & 0 \\ 0 & -k_3 & \ddots & -k_{n-1} \\ 0 & 0 & -k_{n-1} & k_n \end{bmatrix} \tag{5.51c}$$

we can then write the energy terms (Equation 5.50) more compactly as

$$T_{ser} = \frac{1}{2}\dot{\mathbf{X}}^T \mathbf{M}\dot{\mathbf{X}} \quad \text{and} \quad V_{ser} = \frac{1}{2}\mathbf{X}^T \mathbf{K}\mathbf{X} \tag{5.52}$$

To make the same argument about equating the peak potential (stored) and kinetic energies, we note that these peaks are equal for those natural frequencies that represent eigenvalues for the equations of motion for this series system of oscillators

$$M\ddot{X} + KX = 0 \tag{5.53}$$

and those eigenvalues are the solutions of

$$\left[-\omega_i^2 M + K\right]X_i = 0 \tag{5.54}$$

Thus, the eigenvalues are then found from the Rayleigh quotient

$$R_{ser}(X_i) = \omega_i^2 = \frac{X_i^T K X_i}{X_i^T M X_i} \tag{5.55}$$

We then see that each natural frequency (or its square) ω_i^2 is a function of its corresponding eigenvector X_i. If we knew these eigenvectors, which we can find by solving the eigenvalue problem (Equation 5.54), we could calculate the natural frequencies. But (and here is where we get to the highlight) can we estimate the fundamental frequency using the Rayleigh quotient without having already determined the X_i?

So let us follow Rayleigh's lead and imagine that we could approximate the eigenvalues we seek by generalizing Equation (5.55) for an arbitrary displacement vector input X; that is, imagine the Rayleigh quotient as

$$R_{ser}(X) = \frac{X^T K X}{X^T M X} \tag{5.56}$$

This defines an infinite number of frequencies, depending on the value of the displacement vector X.

We can determine that fundamental frequency: We start by recalling that (1) each eigenvector X_i is known only to within an unknown, arbitrary amplitude, and (2) the eigenvectors form an orthogonal set. In fact, they are *mass orthogonal*—meaning that we write the orthogonality condition as

$$X_j^T M X_i = 0 \quad \text{for} \quad i \neq j \tag{5.57}$$

We also impose a normality condition just to make the following matrix arithmetic easier. In particular, we stipulate that

$$X_j^T M X_i = 1 \quad \text{for} \quad i = j \tag{5.58}$$

Again, while the orthogonality condition (Equation 5.57) is an inherent part of the eigenvalue process, we impose the normality condition (Equation 5.58) for convenience. Also, we note that because of the normality condition, all of the natural frequencies found from the Rayleigh quotient (Equation 5.55) are

$$R(\mathbf{X}_i) = \omega_i^2 = \mathbf{X}_i^T \mathbf{K} \mathbf{X}_i \qquad (5.59)$$

Now to the main point: We want to use the Rayleigh quotient (Equation 5.56) to get estimates of a system's fundamental frequency without having determined the actual eigenvalues. Put another way, we want to get the fundamental frequency "for free"—or at least cheaply—without incurring the expense of actually solving Equation (5.54) for the eigenvectors. We start by assuming that we know enough about the physics of the problem we are modeling that we can construct an approximate eigenvector, or *trial solution,* that can be represented in the abstract in terms of the unknown eigenvectors:

$$\tilde{\mathbf{X}} = \sum_{i=1}^{n} p_i \mathbf{X}_i \qquad (5.60)$$

Since we are interested in the fundamental, or lowest, natural frequency, we are really trying to approximate or imitate the presumably unknown first eigenvector $\mathbf{X}_1$. Thus, we will assume that in our approximation $p_1 = 1$ and that the remaining coefficients are small; that is, $p_j \ll 1$ for $j = 2, 3,...n$. We now substitute our approximation (Equation 5.60) into our Rayleigh quotient (Equation 5.56) and take into account both the orthogonality (Equation 5.57) and normality (Equation 5.58) conditions to find that

$$R_{ser}(\tilde{\mathbf{X}}) = \frac{\left(\sum_{i=1}^{n} p_i \mathbf{X}_i^T\right) \mathbf{K} \left(\sum_{i=1}^{n} p_i \mathbf{X}_i\right)}{\left(\sum_{i=1}^{n} p_i \mathbf{X}_i^T\right) \mathbf{M} \left(\sum_{i=1}^{n} p_i \mathbf{X}_i\right)} = \frac{\sum_{i=1}^{n} \left[p_i^2 \left(\mathbf{X}_i^T \mathbf{K} \mathbf{X}_i\right)\right]}{\sum_{i=1}^{n} \left[p_i^2 \left(\mathbf{X}_i^T \mathbf{M} \mathbf{X}_i\right)\right]} = \frac{\omega_1^2 + \sum_{i=2}^{n} p_i^2 \omega_i^2}{1 + \sum_{i=2}^{n} p_i^2} \qquad (5.61)$$

Because we have assumed that the coefficients p_j are small, it is clear that we can approximate the denominator in the quotient (Equation 5.61) with a binomial expansion so that the quotient becomes

$$R_{ser}(\tilde{\mathbf{X}}) \cong \omega_1^2 + \sum_{i=2}^{n} p_i^2 \omega_i^2 \left(1 - \sum_{i=2}^{n} p_i^2\right) = \omega_1^2 + \sum_{i=2}^{n} \left[p_i^2 \left(\omega_i^2 - \omega_1^2\right)\right] \qquad (5.62)$$

Or, finally, we see that if we choose a trial function (Equation 5.60) carefully, we approximate the fundamental frequency quite well; that is, we get

$$R_{ser}(\tilde{\mathbf{X}}) = \omega_1^2 + O(p^2) \tag{5.63}$$

Since we took the p_j to be small, we see here that the corrections to ω_1^2 are really small—in fact, quite negligible!

Two last notes before we leave this demonstration of Rayleigh's quotient. First, the approximations (Equation 5.62 or 5.63) are greater than ω_1^2 itself; that is, our Rayleigh estimates bound their exact counterparts from above. We see this clearly in Equation (5.62) because $\omega_1^2 - \omega_j^2 > 0$. This is in accord with an intuition that says the deflections corresponding to a trial function are more constrained than the exact eigenfunction $\mathbf{X}_i$, thus suggesting a stiffer system; therefore, we would expect frequency estimates to be higher than the exact frequencies. Second, we organized our Rayleigh estimate specifically to approximate the *lowest* natural frequency (i.e., the fundamental frequency). We could have generalized this (as is generally done in the literature) to approximate the jth natural frequency, where $1 < j < n$, but we chose to focus our analysis on our primary interest.

Conclusions

In this chapter we developed extensions of the well-known Castigliano theorems in the context of modeling the behavior of coupled discrete systems. We showed that the proper apportionment of loads among parallel system elements could be determined by minimizing the displacement of that parallel system as dictated by Castigliano's second theorem. Similarly, we demonstrated that we could find the correct distribution of the displacements of a set of series elements by minimizing the load in that series system as defined in Castigliano's first theorem. These extensions will provide a means for apportioning loads in coupled continuous systems, as we will see in Chapter 6, where we examine coupled Timoshenko beams supporting discrete and continuous loads.

We also introduced Rayleigh's quotient as a means to estimate the fundamental frequencies of coupled discrete series systems, showing that it produced approximations that bounded the exact frequencies from above.

Bibliography

Bishop, R. E. D., and D. C. Johnson. 1960. *The mechanics of vibration.* Cambridge, England: Cambridge University Press.

Dym, C. L. 1997. *Structural modeling and analysis.* New York: Cambridge University Press.
———. 2010. Extending Castigliano's theorems to model the behavior of coupled systems. *Journal of Applied Mechanics* 77 (6): 061005-1-6.
Dym, C. L., and I. H. Shames. 1973. *Variational methods in solid mechanics.* New York: McGraw-Hill.
Dym, H. 2006. *Linear algebra in action.* Providence, RI: American Mathematical Society.
Temple, G., and W. G. Bickley. 1956. *Rayleigh's principle and its applications to engineering.* Mineola, NY: Dover Publications.
Tongue, B. H. 2002. *Principles of vibration.* New York: Oxford University Press.

6

Buildings Modeling as Coupled Beams: Static Response Estimates

Summary

In this chapter we present both exact solutions and analytical estimates of the deflections of a pair of coupled Timoshenko beams as an instance of modeling a coupled continuous system. The coupled system itself is an attempt to extend current attempts to model entire buildings as either a single beam or a coupled pair of beams. Our exact solutions are derived by solving the Timoshenko beam equations for the case where two beams are coupled in parallel, for two different support sets, and for several transverse loads. We derive analytical estimates of the same deflections using the extension of Castigliano's theorem for coupled systems that we derived in Chapter 5. We test our deflection estimates against their corresponding exact solutions and comment on the apportionment of loads between the coupled beams.

Introduction

Simple "back of the envelope" models are always useful in engineering, and "neat" formulas are often admired as much for their simplicity as for their accuracy and utility. The work described here was originally motivated by a desire to provide simple, accurate estimators for the deflections of tall buildings. A lot of work has been done on modeling buildings in terms of beam components. Suffice it to say that the individual beams used to construct such building models include elementary Euler–Bernoulli beams, shear beams, and Timoshenko beams. These three beam models are *series* beam models (see Figure 6.1) in which (1) bending and shear deflections add, (2) bending and shear stiffnesses add reciprocally, and (3) each component carries the same common load(s).

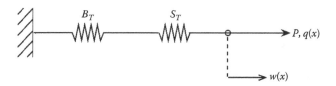

FIGURE 6.1
A discrete model of a Timoshenko beam, reflecting bending B_T and shear S_T elements responding in series. Euler beams follow by allowing $S_T \to \infty$ and shear beams from $B_T \to \infty$.

However, while the shear and bending deflections of individual beams add in *series*, the aggregation of these beam elements or components in a building model requires a *parallel* formulation (Figure 6.2). The distinction between series and parallel formulations is more evident in models of the dynamic response of tall buildings because empirical data and practical experience show the natural frequencies of tall buildings varying as $1/L$, while the fundamental frequencies of both elementary and Timoshenko beams vary as $1/L^2$. On the other hand, when we couple an Euler beam in *parallel* to a shear beam by requiring a common deflection (see Figure 6.3), we find a fundamental frequency that varies as $1/L$ over a broad range of interest. Further, this coupled Euler-to-shear beam has successfully modeled the dynamic response of a building composed of a shear core and a tube.

We also point out that new, framed structures are often added to existing structures to create a parallel load path designed to improve the structure's seismic response (see Figure 6.4). Here the parallel modeling of components is clear.

We would also note coupled Timoshenko beams have been useful for modeling the dynamic response of multiwalled carbon nanotubes, so the kinds of models we describe here can be applied to continuous coupled elastic systems beyond building models.

We will provide both exact and approximate solutions for a system that couples two Timoshenko beams in parallel, with each Timoshenko beam

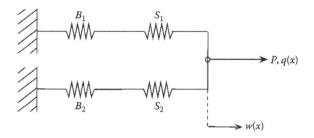

FIGURE 6.2
A discrete model of two Timoshenko beams, each reflecting bending B_i and shear S_i elements responding in series, coupled together in a parallel system.

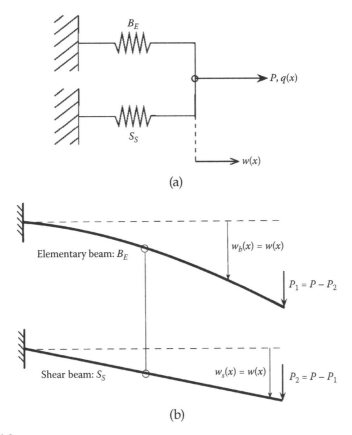

FIGURE 6.3
Discrete and continuous models of an Euler beam in bending B_E coupled in parallel with a pure shear S_S element.

supporting *both* bending and shear deformation in series (Figure 6.2). This is a logical extension of the parallel coupling of an Euler beam to a shear beam described earlier, where now neither component is "pure": The (formerly pure) Euler beam element also supports shear deformation and the (formerly pure) shear beam element supports bending. The components' total deflections are constrained to be equal one to another (and thus to the total deflection), the bending and shear stiffnesses of each Timoshenko component add linearly, and the external load is distributed among the various components in proportion to each component's stiffness as a fraction of the total stiffness.

While we can easily apportion the external load(s) among the parallel components for discrete parallel models wherein the external load is a single "concentrated" load, it is not clear how to estimate such load apportionment when the load is spatially distributed over a beam model's length. However,

FIGURE 6.4
The remodeled San Diego Medical Center of the US Department of Veteran Affairs is an instance where an added parallel structure increased a building's ability to respond to seismic events (designed by Leo A. Daly Associates). In this case, external stairways were replaced with new ones connected by external frames that provided an additional "cage" to inhibit lateral movement. (Photo by permission of Tom Bonner Photography.)

we will see that the load apportionment is determined naturally as part of obtaining deflection estimates, and that the effective stiffness of a model that couples two Timoshenko beams in parallel depends on three dimensionless ratios that we will fully detail later:

- Two, $\lambda_i^2 = S_i L^2 / B_i$ ($i = 1, 2$), relate the pairs of bending and shear stiffnesses that subsume four material-five geometric properties of the two beams.
- A third, $a^2 = \bar{V} L^2 / \bar{M}$, reflects weighted averages of the given applied load.

The numerical values of interest for the building parameters λ_i are extrapolated from a parameter α used in the aforementioned modeling of the dynamic response of coupled Euler–shear cantilever systems (see also Chapter 7): $0 \le \alpha \le 1.5$ for shear wall and braced frame buildings, $1.5 \le \alpha \le 5$ for *dual-system* buildings combining moment-resisting frames with shear walls or braced frames, and $5 \le \alpha \le 20$ for moment-resisting frames.

In Equation (6.16) we will define our version of α, called α_{12}, as a special case derived from a pair of Timoshenko beams coupled together in parallel. The external loads we consider include a tip load on a cantilever, as well as loads distributed uniformly and exponentially along the beam length.

We will see that our deflection estimates for the coupled beam systems are quite accurate when we compare them to their exact counterparts, and these estimates are easily and meaningfully applied.

Coupled Timoshenko Beams: Exact Solutions

We now consider two Timoshenko cantilevers connected in parallel (Figure 6.5). Each beam shares a common total deflection $w_i(x) = w(x)$ ($i = 1, 2$), has its own bending rotation (or Euler angle) $\psi_i(x)$, and has its own bending and shear stiffnesses, $E_i I_i$ and $k_i G_i A_i$. There are four constitutive relations, each an extension of the constitutive relations of a single Timoshenko beam: two moment-curvature relations,

$$M_i(x) = -B_i \frac{d\psi_i(x)}{dx} \tag{6.1}$$

and two relations between shear resultants and their net shear angles,

$$V_i(x) = S_i \left(\frac{dw(x)}{dx} - \psi_i(x) \right) \tag{6.2}$$

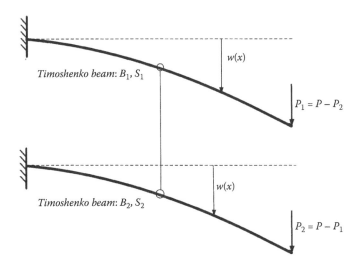

FIGURE 6.5
A system of two Timoshenko cantilever beams, each with bending stiffness B_i coupled in parallel to the shear stiffness S_i.

We find the governing equations for this pair of Timoshenko beams connected in parallel by minimizing the total potential energy, Π:

$$\Pi = \frac{B_1}{2} \int_0^L \left(\frac{d\psi_1(x)}{dx} \right)^2 dx + \frac{S_1}{2} \int_0^L \left(\frac{dw(x)}{dx} - \psi_1(x) \right)^2 dx$$

$$+ \frac{B_2}{2} \int_0^L \left(\frac{d\psi_2(x)}{dx} \right)^2 dx + \frac{S_2}{2} \int_0^L \left(\frac{dw(x)}{dx} - \psi_2(x) \right)^2 dx \qquad (6.3)$$

$$- \int_0^L q(x)w(x)\,dx - P^*w(L) + M_1^*\psi_1(x) + M_2^*\psi_2(x)$$

We see that Equation (6.3) also reflects the facts that the beam system supports a distributed load $q(x)$, the total tip load is $P^* = P_1^* + P_2^*$, and two separate tip moments (M_i^*) are prescribed because each beam exhibits its own bending angle $\psi_i(x)$. Then, if we take the first variation of Equation (6.3) and set it to zero, we get three equations of equilibrium, which we write in terms of stress resultants:

$$\frac{dM_1(x)}{dx} - V_1(x) = 0$$

$$\frac{dM_2(x)}{dx} - V_2(x) = 0 \qquad (6.4)$$

$$\frac{dV_1(x)}{dx} + \frac{dV_2(x)}{dx} + q(x) = 0$$

We can also express these equations of equilibrium in terms of displacements as

$$B_1 \frac{d^2\psi_1(x)}{dx^2} + S_1 \left(\frac{dw(x)}{dx} - \psi_1(x) \right) = 0$$

$$B_2 \frac{d^2\psi_2(x)}{dx^2} + S_2 \left(\frac{dw(x)}{dx} - \psi_2(x) \right) = 0 \qquad (6.5)$$

$$(S_1 + S_2) \frac{d^2w(x)}{dx^2} - S_1 \frac{d\psi_1(x)}{dx} - S_2 \frac{d\psi_2(x)}{dx} + q(x) = 0$$

Equation (6.5) comprises a sixth-order system that requires the satisfaction of six corresponding boundary conditions chosen from the following six duals:

Either	$M_1(0) = 0$	or	$\delta\psi_1(0) = 0$	(6.6a)
Either	$M_2(0) = 0$	or	$\delta\psi_2(0) = 0$	(6.6b)
Either	$V_1(0) + V_2(0) = 0$	or	$\delta w(0) = 0$	(6.6c)
Either	$M_1(L) - M_1^* = 0$	or	$\delta\psi_1(L) = 0$	(6.6d)
Either	$M_2(L) - M_2^* = 0$	or	$\delta\psi_2(L) = 0$	(6.6e)
Either	$V_1(L) + V_2(L) - P = 0$	or	$\delta w(L) = 0$	(6.6f)

The solutions of the equilibrium equations (Equation 6.5), subject to the boundary conditions (Equations 6.6a–6.6f), depend on the precise nature of the distributed load $q(x)$. For a uniformly distributed load $q(x) = q_0$ on the coupled system, we find the resulting transverse displacement (see Appendix A for the straightforward, if tedious, details, including the values of the constants C_i):

$$w_{q_0/exact}(x) = \left(\frac{q_0 L^4}{B}\right)\left[\begin{array}{l} C_0 + C_1\left(\frac{x}{L}\right) + C_2\left(\frac{x}{L}\right)^2 + C_3\left(\frac{x}{L}\right)^3 \\ + C_4 \sinh\frac{\gamma x}{L} + C_5 \cosh\frac{\gamma x}{L} + \frac{1}{24}\left(\frac{x}{L}\right)^4 \end{array}\right] \quad (6.7a)$$

and the corresponding bending rotations (or Euler angles), which take the same general form for both of the Timoshenko beams, as

$$\psi_{i/q_0}(x) = \left(\frac{q_0 L^3}{B}\right)\left[\begin{array}{l} \left(C_1 + \frac{6}{\lambda_i^2}C_3\right) + \left(2C_2 + \frac{1}{\lambda_i^2}\right)\left(\frac{x}{L}\right) + 3C_3\left(\frac{x}{L}\right)^2 + \frac{1}{6}\left(\frac{x}{L}\right)^3 \\ + \frac{\gamma\lambda_i^2}{\lambda_i^2 - \gamma^2}\left(C_4 \cosh\frac{\gamma x}{L} + C_5 \sinh\frac{\gamma x}{L}\right) \end{array}\right] \quad (6.7b)$$

We have introduced several parameters into Equations (6.7a) and (6.7b), including sums of the bending and shear moduli,

$$B = B_1 + B_2 \quad \text{and} \quad S = S_1 + S_2 \quad (6.8)$$

and three dimensionless parameters,

$$\lambda_i^2 = \frac{S_i L^2}{B_i}, \quad \lambda^2 = \frac{SL^2}{B} = \frac{(S_1 + S_2)L^2}{(B_1 + B_2)} \tag{6.9a}$$

as well as a combination of those three dimensionless parameters,

$$\gamma^2 = \frac{\lambda_1^2 \lambda_2^2}{\lambda^2} = L^2 \frac{(B_1 + B_2)}{(S_1 + S_2)}\left(\frac{S_1}{B_1}\right)\left(\frac{S_2}{B_2}\right) \tag{6.9b}$$

We determine the six independent constants in Equations (6.7a) and (6.7b) by satisfying those duals from Equations (6.6a)–(6.6f) that correspond to the problem we are modeling. For a cantilever system supported at $x = 0$ and free at $x = L$, the appropriate boundary conditions expressed in displacement components are

$$w(0) = 0 \qquad \psi_1(0) = 0 \qquad \psi_2(0) = 0 \tag{6.10a}$$

$$\frac{d\psi_1(L)}{dx} = 0 \qquad \frac{d\psi_2(L)}{dx} = 0 \qquad S\frac{dw(L)}{dx} - (S_1\psi_1(L) + S_2\psi_2(L)) = 0 \quad (6.10b)$$

We show the details of the complete solution in Appendix A, but it is worth noting that the tip deflection, which might characterize a building's *drift*, is

$$w_{q0/exact}(L) = \left(\frac{q_0 L^4}{B}\right)\left[\frac{1}{8} - \frac{1}{2\gamma^2} + \frac{\lambda_1^2 + \lambda_2^2}{2\lambda_1^2\lambda_2^2} - \frac{(\gamma^2 - \lambda_1^2)(\gamma^2 - \lambda_2^2)}{\gamma^4\lambda_1^2\lambda_2^2}\left(1 - \frac{1 + \gamma\sinh\gamma}{\cosh\gamma}\right)\right] \tag{6.11}$$

Note that Equation (6.11) is appropriately symmetric with respect to beams 1 and 2; that is, the result is indifferent to which beam is "on top." We also see from Equation (6.11) that when the two beams have equal material-geometry ratios (i.e., $\lambda_1^2 = \lambda_2^2 = \lambda_c^2$), it then follows that $\gamma^2 = \lambda_c^2$. In this critical case, by examining Equation (6.11) we can show that

$$w_{q0/exact}(L) = \left(\frac{q_0 L^4}{B}\right)\left[\frac{1}{8} + \frac{1}{2\lambda_c^2}\right] \tag{6.12}$$

Another special case of interest (recall our discussion in the Introduction) occurs when an Euler beam ($i = 1$) is coupled to a shear beam ($i = 2$). When subject to a uniform load, the equilibrium equations (Equation 6.5) for this configuration reduce to a single one:

$$B_1 \frac{d^4 w(x)}{dx^4} - S_2 \frac{d^2 w(x)}{dx^2} = q_0 \tag{6.13}$$

that is subject to the boundary conditions

$$w(0) = 0 \qquad\qquad w'(0) = 0 \tag{6.14a}$$

$$B_1 w'''(L) - S_2 w'(L) = 0 \qquad w''(L) = 0 \tag{6.14b}$$

We then solve the differential Equation (6.13) and satisfy the boundary conditions (Equations 6.14a and 6.14b) to find the tip deflection as

$$w_{q_0/exact/E-S}(L) = \frac{q_0 L^2}{\alpha_{12}^2 S_2 \cosh\alpha_{12}} \left[\left(1 + \frac{\alpha_{12}^2}{2} \right) \cosh\alpha_{12} - (1 + \alpha_{12}\sinh\alpha_{12}) \right] \tag{6.15}$$

Here, α_{12} is a dimensionless "mixed" stiffness constant, our version of the parameter α discussed earlier:

$$\alpha_{12}^2 \triangleq S_2 L^2 / B_1 \tag{6.16}$$

By judiciously applying small angle approximations, when the dimensionless quantity $\alpha_{12} \ll 1$, we can show that

$$w_{q_0/exact/E-S}(L)|_{\alpha_{12} \ll 1} \Rightarrow \frac{q_0 L^4}{8 B_1} \tag{6.17}$$

Equation (6.17) agrees exactly with the limit of the exact tip deflection (Equation 6.11) for small values $\alpha_{12} \ll 1$. Similarly, for large values $\alpha_{12} \gg 1$, it follows from Equation (6.15) that

$$w_{q_0/exact/E-S}(L)|_{\alpha_{12} \gg 1} \Rightarrow \frac{q_0 L^2}{2 S_2} \tag{6.18}$$

It is also interesting to note that the exact solution (Equation 6.11) produces the same limiting value of Equation (6.15), and we show the details of that in Appendix B. Moreover, we will show in the next section that our Castigliano-based estimates of the deflection will produce the same limiting values for both small and large values of α_{12}.

Consider now also the limit of Equation (6.11) when the two coupled beams are both Euler beams. In this instance, $S_i \to \infty$ and $\lambda_i^2 \to \infty$, while $\lambda_i^2/S_i \to L^2/B_i$, so that we find that

$$w_{q(x)/2E}(L) = \frac{q_0 L^4}{8B} \tag{6.19}$$

We note that Equations (6.17) and (6.19) also agree with the exact tip deflection of an Euler cantilever supporting a uniform load.

Coupled Timoshenko Beams: Castigliano Estimates (I)

We now develop an estimate for the tip deflection (or building drift) of such a set of coupled Timoshenko cantilevers by applying our extension of Castigliano's theorems of Chapter 5 to this continuous system. Remember that these theorems are a way to discretize continuous systems to find forces (first theorem) or displacements (second theorem) at specific points in those systems. We write the complementary energy for n Timoshenko beams to incorporate both bending and shear energies in each beam:

$$U_{n-Time}^* = \sum_{i=1}^{n}\left[U_{bi}^* + U_{si}^*\right] = \sum_{i=1}^{n}\left[\int_0^L \frac{(M_i(x))^2}{2B_i}dx + \int_0^L \frac{(V_i(x))^2}{2S_i}dx\right] \tag{6.20}$$

We get the tip deflection from Castigliano's second theorem by minimizing the complementary energy (Equation 6.20) with respect to the load P:

$$w_{q(x)}(L) = \left.\frac{dU_{n-Time}^*}{dP}\right|_{P\to 0} = \sum_{i=1}^{n}\left[\int_0^L \frac{V_i(x)}{S_i}\frac{dV_i(x)}{dP}dx + \int_0^L \frac{M_i(x)}{B_i}\frac{dM_i(x)}{dP}dx\right] \tag{6.21}$$

Now we must identify the shear forces and moments in each component beam. For the system, the total shear and moments for the cantilevers are given by

$$V(x) = -\int_H^x q(\xi)d\xi$$

$$M(x) = \int_H^x V(\eta)d\eta = -\int_H^x\int_H^\eta q(\xi)d\xi\,d\eta \tag{6.22}$$

Note that Equations (6.20)–(6.22) are valid for a variety of beams, including Euler bending, shear, and Timoshenko beams.

We write the total load applied to the cantilever system to account for both a distributed load and a tip load as

$$q(x) \Rightarrow q(x) + P\delta_D(x-L) = \sum_{i=1}^{n} (q_i(x) + P_i\delta_D(x-L)) \qquad (6.23)$$

Remember that we can also use the tip load P as a phantom load (together with the Dirac spike $\delta_D(x-L)$) to determine the cantilever system's tip deflection for any distributed load $q(x)$. Taking heed of what we saw in Chapter 5, we now assume that we can linearly apportion or allocate both the distributed and tip loads between n Timoshenko beams, with dimensionless load apportionment ratios r_i, such that

$$(q_i(x) + P_i\delta_D(x-L)) = r_i(q(x) + P\delta_D(x-L)) \qquad (6.24)$$

In view of Equations (6.22)–(6.24), we can write the moment and shear resultants in each Timoshenko beam as

$$V_i(x) = r_i\left[-\int_0^x q(\xi)\,d\xi + P\right] \equiv r_i(V(x) + P)$$

$$(6.25)$$

$$M_i(x) = r_i\left[-\int_0^x \int_0^\eta q(\xi)\,d\xi\,d\eta + P(x-L)\right] \equiv r_i(M(x) + P(x-L))$$

We can identify the derivatives $dV_i(x)/dP = r_i$ and $dM_i(x)/dP = r_i(x-L)$ from Equation (6.25), so we can write the tip deflection (Equation 6.21) as

$$w_{q(x)}(L) = \sum_{i=1}^{n}\left[\left(\frac{r_i^2}{S_i}\right)\left(\int_0^L V(x)\,dx\right) + \left(\frac{S_i}{B_i}\right)\int_0^L M(x)(x-L)\,dx\right] \qquad (6.26)$$

In the deflection (Equation 6.26) we can identify two constants that represent weighted averages of the shear and moment resultants:

$$\bar{V} = \int_0^L V(x)\,dx \quad \text{and} \quad \bar{M} = \int_0^L M(x)(x-L)\,dx \qquad (6.27)$$

We now also introduce into Equation (6.26) a pair of dimensionless ratios defined for each beam *individually*:

$$\lambda_i^2 = \frac{L^2 S_i}{B_i} = \frac{k_i G_i}{E_i}\left(\frac{A_i L^2}{I_i}\right) \tag{6.28}$$

We might be tempted to conclude from the second form of Equation (6.28) that each of the λ_i^2 will be a relatively large number because the dimensionless ratio $I_i/A_i L_i^2$ is the ith beam's *slenderness ratio*, which is usually small. However, we ought to keep in mind that the shear and Young's moduli of each Timoshenko are independent of one another and thus may have very different magnitudes. With the aid of Equations (6.27) and (6.28) we can write the tip deflection (Equation 6.26) of the cantilever system as

$$w_{q(x)}(L) = \sum_{i=1}^{n}\left[r_i^2\left(\frac{\bar{V}}{S_i}\right)\left(1+\lambda_i^2\left(\frac{\bar{M}}{L^2\bar{V}}\right)\right)\right] = \sum_{i=1}^{n}\left[r_i^2\left(\frac{\bar{V}}{S_i}\right)\left(1+\frac{\lambda_i^2}{a^2}\right)\right] \tag{6.29}$$

in which we have defined our last dimensionless constant as a ratio of the weighted shear force and moment resultants,

$$a^2 = \frac{L^2\bar{V}}{\bar{M}} \tag{6.30}$$

We should also note that Equation (6.29) can be cast so that its driver is the weighted stress resultant $\bar{M}$ rather than $\bar{V}$; that is,

$$w_{q(x)}(L) = \sum_{i=1}^{n}\left[r_i^2\left(\frac{\bar{M}}{B_i}\right)\left(1+\frac{a^2}{\lambda_i^2}\right)\right] \tag{6.31}$$

Now, to relate this analysis to the discrete formulation of Chapter 5 more strongly, we note that either of the deflections (Equation (6.29) or (6.31)) can be cast in terms of an effective stiffness for each Timoshenko beam appropriate to either of the weighted resultants:

$$K_{T\bar{V}i} = \frac{S_i}{1+\lambda_i^2/a^2} \quad \text{or} \quad K_{T\bar{M}i} = \frac{B_i}{1+a^2/\lambda_i^2} \tag{6.32}$$

Then, in view of Equation (6.32), we see the midpoint deflection in two forms:

$$w_{q(x)}(L) = \sum_{i=1}^{n} r_i^2\left(\frac{\bar{V}}{K_{T\bar{V}i}}\right) = \sum_{i=1}^{n} r_i^2\left(\frac{\bar{M}}{K_{T\bar{M}i}}\right) \tag{6.33}$$

Equation (6.33)—which contains direct counterparts of Equation (5.11)—represents two forms of the discretized tip deflection of a system of n Timoshenko cantilevers subjected to the distributed load that produced the weighted shear $\bar{V}$ and the weighted moment $\bar{M}$.

To focus on just two coupled Timoshenko beams, we note that there is only one independent apportionment factor we need to consider:

$$r_1 = r \quad \text{and} \quad r_2 = (1 - r) \tag{6.34}$$

Then, the first formulation in Equation (6.33) becomes

$$w_{q(x)}(L) = \left(\frac{r^2 \bar{V}}{K_{T\bar{V}1}} \right) + \left(\frac{(1-r)^2 \bar{V}}{K_{T\bar{V}2}} \right) \tag{6.35}$$

Now, as articulated in Chapter 5 for discrete systems, we minimize Equation (6.35) with respect to the apportionment factor r to find the value of that factor that minimizes the tip deflection of the two-beam system. This, in turn, gives us a compatible deflection for the system that properly preserves equilibrium of the applied loading with the forces carried in the two Timoshenko beams. In this case we easily find the minimizing value of the apportionment factor to be

$$r_{\min} = \frac{K_{T\bar{V}1}}{K_{T\bar{V}1} + K_{T\bar{V}2}} = \frac{1}{1 + \frac{K_{T\bar{V}2}}{K_{T\bar{V}1}}} = \frac{1}{1 + \frac{1+\lambda_1^2/a^2}{1+\lambda_2^2/a^2} \left(\frac{S_2}{S_1} \right)} \tag{6.36}$$

and we find the corresponding tip deflection as

$$w_{q(x)}(L) = \frac{\bar{V}}{K_{T\bar{V}1} + K_{T\bar{V}2}} \tag{6.37}$$

We can manipulate the deflection (Equation 6.37) to express it in terms of the respective shear stiffnesses,

$$w_{q(x)}(L) = \frac{\bar{V}}{\frac{S_1}{1+\lambda_1^2/a^2} + \frac{S_2}{1+\lambda_2^2/a^2}} \tag{6.38a}$$

or in terms of the bending stiffnesses,

$$w_{q(x)}(L) = \frac{\bar{M}}{\frac{B_1}{1+a^2/\lambda_1^2} + \frac{B_2}{1+a^2/\lambda_2^2}} \tag{6.38b}$$

Note also that the deflection formulas (Equations 6.37, 6.38a, and 6.38b) are completely (and properly) *symmetric* with respect to beams 1 and 2.

We can now derive special cases, depending on whether our interest is in a given loading or in a particular system configuration. For example, for a tip load on a Timoshenko cantilever system, the shear and moment resultants are the familiar $V(x) = P$ and $M(x) = P(x - L)$, so the weighted shear and moment resultants can be found to be

$$\bar{V} = PL, \quad \bar{M} = \frac{PL^3}{3}, \quad a^2 = 3 \tag{6.39}$$

The corresponding tip deflection is then

$$w_P(L) = \frac{PL}{\dfrac{S_1}{1+\lambda_1^2/3} + \dfrac{S_2}{1+\lambda_2^2/3}} = \frac{PL^3/3}{\dfrac{B_1}{1+3/\lambda_1^2} + \dfrac{B_2}{1+3/\lambda_2^2}} \tag{6.40}$$

and the load apportionment ratio is

$$r_{1/P} = \frac{\dfrac{S_1}{1+\lambda_1^2/3}}{\dfrac{S_1}{1+\lambda_1^2/3} + \dfrac{S_2}{1+\lambda_2^2/3}} = \frac{\dfrac{B_1}{1+3/\lambda_1^2}}{\dfrac{B_1}{1+3/\lambda_1^2} + \dfrac{B_2}{1+3/\lambda_2^2}} \tag{6.41}$$

These results correspond exactly to the discrete result, which we would expect for a tip load on the cantilever system.

To illustrate the effect of changes in the beams being modeled, we stay with a tip load but consider an Euler beam (beam 1; $S_1 \to \infty$, $\lambda_1^2 \to \infty$) coupled to a shear beam (beam 2; $B_2 \to \infty$, $\lambda_2^2 \to 0$), for which the first apportionment ratio is

$$r_{1/P} = \frac{3B_1/L^3}{3B_1/L^3 + S_2/L} \equiv \frac{1}{1 + S_2L^2/3B_1} \tag{6.42}$$

while the concomitant tip deflection (cf. Equations 6.38a and 6.38b) is

$$w_P(L) = \frac{P}{3B_1/L^3 + S_2/L} \equiv \frac{PL^3/3B_1}{1 + S_2L^2/3B_1} \tag{6.43}$$

Equations (6.42) and (6.43) clearly reflect the parallel stiffnesses of the Euler ($a^2B_1/H^3 = 3B_1/H^3$) and shear (S_2/H) beams.

We now turn to a more interesting special case (identified earlier as a *critical* case) that occurs when the two dimensionless beam material-geometry ratios are the same (i.e., $\lambda_1^2 = \lambda_2^2 \triangleq \lambda_c^2$). Then, for any load $q(x)$, Equations (6.38a) and (6.38b) become

$$w_{q(x)/crit}(L) = \left(1 + \lambda_c^2/a^2\right)\left(\frac{\overline{V}}{S_1 + S_2}\right) \equiv \left(1 + a^2/\lambda_c^2\right)\left(\frac{\overline{M}}{B_1 + B_2}\right) \qquad (6.44)$$

The corresponding load apportionment ratios are

$$r_{1/crit} = \frac{S_1}{S_1 + S_2} \equiv \frac{B_1}{B_1 + B_2} \quad \text{and} \quad r_{2/crit} = \frac{S_2}{S_1 + S_2} \equiv \frac{B_2}{B_1 + B_2} \qquad (6.45)$$

Note that the tip deflection clearly depends on the critical material-geometry ratio λ_c^2, but the load apportionment factors do not. The results in Equation (6.44) are precisely those obtained with the exact solution and shown in Equation (6.12).

To illustrate how the tip deflection varies with the actual load distribution, we now consider a cantilever subjected to a uniform distributed load—that is, $q(x) = q_0$, a constant. The shear and moment resultants follow from Equation (6.22) as

$$V(x) = -q_0(x - L)$$

$$M(x) = -\frac{q_0(x - L)^2}{2} \qquad (6.46)$$

Then, making use of all the machinery given in Equations (6.20)–(6.33), we can proceed directly to calculate the parameters defined in Equations (6.27) and (6.30):

$$\overline{V} = \frac{q_0 L^2}{2}, \quad \overline{M} = \frac{q_0 L^4}{8}, \quad a^2 = 4 \qquad (6.47)$$

so that the tip deflection of a cantilever under a uniform load is

$$w_{q_0}(L) = \frac{1 + \lambda_1^2/4}{1 + \frac{1+\lambda_1^2/4}{1+\lambda_2^2/4}\left(\frac{S_2}{S_1}\right)}\left(\frac{q_0 L^2}{2S_1}\right) = \frac{1 + 4/\lambda_1^2}{1 + \frac{1+4/\lambda_1^2}{1+4/\lambda_2^2}\left(\frac{B_2}{B_1}\right)}\left(\frac{q_0 L^4}{8B_1}\right) \qquad (6.48)$$

We can easily see from the second result in Equation (6.48) that we get the kinds of limiting cases we have seen before (in Equations 6.17 and 6.19) for the tip deflection of a uniformly loaded cantilever, thus suggesting again that estimates obtained in less well-known cases are also likely to be good. We also see that Equations (6.38a), (6.38b), and (6.36) are easily implemented estimates of the deflection and load apportionment as a function of the constituent beam elements and the factor a. We present numerical comparisons for a variety of values of λ_1^2 and λ_2^2 next to confirm that our simple analytical estimates compare very well with the exact results, for both the deflection and for its attendant load apportionment factor, and we show further special and limiting cases in Appendix C.

Validating Castigliano-Based Deflection Estimates

It is clear from our discussions that the Castigliano-based deflection results are simpler and easier to imagine as back of the envelope tools than their corresponding exact solutions—that is, compare Equations (6.7a) and (6.7b) with Equations (6.38a) and (6.38b). They certainly provide intuitive forms that are easier to manipulate. Consider, for example, the case of a distributed surface load that varies along the length of the cantilever as

$$q(x) = q_n \left(\frac{x}{L} \right)^n$$

(6.49)

Note that each $q_n = q(L)$ represents the peak value of its load, and its physical dimensions are force/length. The case $n = 0$ represents a uniform load (i.e., $q(x) = q_0$). A linear variation ($n = 1$) is used to model seismic input with an equivalent total load $q_1 L/2$. The quadratic case ($n = 2$) is a "smooth" representation of wind loading over the building height, with a resultant $q_2 L/3$. Finally, the limit $n \to \infty$ corresponds to a tip load. Values of the parameters $\bar{V}_n$, $\bar{M}_n$, and a^2, for various values of n are shown in Table 6.1;

TABLE 6.1

Variation of the Parameters $\bar{V}_n$, $\bar{M}_n$, and a_n^2, with n under Loading $q_n(x/L)^n$

n	0	1	2	n	$n \to \infty$
$q_n(x/L)^n$	$1/2$	$1/3$	$1/4$	$1/(n+2)$	$1/n$
$\bar{V}_n/q_n L^2$	$1/8$	$11/120$	$13/180$	$\dfrac{2n^3 + 15n^2 + 31n + 18}{6(n+2)(n^3 + 8n^2 + 19n + 12)}$	$1/3n$
a_n^2	4	$40/11 = 3.64$	$45/13 = 3.46$	$3\left(\dfrac{2n^3 + 16n^2 + 38n + 24}{2n^3 + 15n^2 + 31n + 18} \right)$	$3(1 + 1/2n) \to 3$

these values are easily inserted into Equations (6.38a) and (6.38b) to estimate a cantilever's tip deflection.

Thus, calculating numerical estimates of beam deflections is relatively
easy. But are these estimates any good? They are, as can be confirmed by
comparing the estimates to the exact solution (i.e., Equations 6.7a and 6.7b).
For a uniform load ($n = 0$), the dimensionless Castigliano-based deflection
estimate is, from Equations (6.38a) and (6.38b) and Table 6.1,

$$\frac{Bw_{q_0/estimate}(L)}{q_0 L^4} = \frac{B/2L^2}{\frac{S_1}{1+\lambda_1^2/4} + \frac{S_2}{1+\lambda_2^2/4}} = \frac{B/8}{\frac{B_1}{1+4/\lambda_1^2} + \frac{B_2}{1+4/\lambda_2^2}} \tag{6.50}$$

Table 6.2 displays a few comparisons for ranges of the material-
geometry parameters that are reasonably representative of those that are
used in the modeling of buildings. The estimate of Equation (6.50) appears to
agree rather well with the exact solution found in Equations (6.7a) and (6.7b),
especially in view of the simplicity of that estimate.

If the coupled Timoshenko beams were loaded only at the tip, Equation
(6.40) gives the dimensionless estimate as

$$\frac{Bw_{P/estimate}(L)}{PL^3} = \frac{B/3}{\frac{B_1}{1+3/\lambda_1^2} + \frac{B_2}{1+3/\lambda_2^2}} \tag{6.51}$$

The corresponding exact solution can be shown to be (and we leave the
details for the reader), in analogy with Equations (6.7a) and (6.7b),

$$\frac{Bw_{P/exact}(L)}{PL^3} = \frac{1}{3} + \frac{1}{\lambda_1^2} + \frac{1}{\lambda_2^2} - \frac{1}{\gamma^2} + \frac{(\lambda^2 - \lambda_1^2)(\lambda^2 - \lambda_2^2)}{\lambda^2 \lambda_1^2 \lambda_2^2} \left(\frac{\tanh \gamma}{\gamma}\right) \tag{6.52}$$

TABLE 6.2

Exact Values and Estimates of the Dimensionless Tip
Deflection of a Pair of Coupled Timoshenko Beams Subject
to a Uniform Load for Typical Material-Geometry
Parameters ($\lambda_1^2 = 100.0$, $\lambda_2^2 = 1.0$)

B_2/B_1	$\alpha_{12} = \lambda_2\sqrt{\frac{B_2}{B_1}}$	$\dfrac{Bw_{q_0/exact}(L)}{q_0 L^4}$	$\dfrac{Bw_{q_0/estimate}(L)}{q_0 L^4}$	Error (%)
1.0	1.0	0.2035	0.2152	5.7
2.0	1.41	0.2518	0.2754	9.4
3.0	1.73	0.2864	0.23202	11.8

TABLE 6.3

Exact Values and Estimates of the Dimensionless Tip
Deflection of a Pair of Coupled Timoshenko Beams
Subject to a Tip Load P for Typical Material-Geometry
Parameters ($\lambda_1^2 = 100.0$, $\lambda_2^2 = 1.0$)

B_2/B_1	$\alpha_{12} = \lambda_2\sqrt{\dfrac{B_2}{B_1}}$	$\dfrac{Bw_{P/exact}(L)}{PL^3}$	$\dfrac{Bw_{P/estimate}(L)}{PL^3}$	Error (%)
1.0	1.0	0.5325	0.5461	2.6
2.0	1.41	0.6533	0.6799	4.1
3.0	1.73	0.7375	0.7748	5.1

While Equations (6.51) and (6.52) do not seem to resemble one another, the
numerical results we show in Table 6.3 demonstrate that they produce virtually identical results.

When a uniform load q_0 is applied to an Euler beam (beam 1; $S_1 \to \infty$, $\lambda_1^2 \to \infty$)
that is coupled to a shear beam (beam 2; $B_2 \to \infty$, $\lambda_2^2 \to 0$), an estimate of the tip
deflection may be obtained from Equation (6.50) as

$$\frac{B_1 w_{q0/estimate/E-S}(L)}{q_0 L^4} = \frac{B_1/2L^3}{4B_1/L^3 + S_2/L} = \frac{1}{8 + 2\alpha_{12}^2} \tag{6.53}$$

The exact tip deflection follows from Equation (6.15) as

$$\frac{B_1 w_{q0/exact/E-S}(H)}{q_0 H^4} = \frac{1}{\alpha_{12}^4 \cosh\alpha_{12}}\left[\left(1+\frac{\alpha_{12}^2}{2}\right)\cosh\alpha_{12} - (1+\alpha_{12}\sinh\alpha_{12})\right] \tag{6.54}$$

The resemblance between Equations (6.53) and (6.54) is once again not terribly obvious, yet Equation (6.53) produces precisely the same limiting results
given for small α_{12} (Equation 6.17) and large α_{12} (Equation 6.18). In addition,
the numerical results we give in Table 6.4 show that Equations (6.53) and
(6.54) produce reasonably close results.

When the system of an Euler beam coupled to a shear beam is subjected
to a tip load P, the tip deflection estimate that follows from Equation (6.43)
or (6.51) is

$$\frac{B_1 w_{P/estimate/E-S}(H)}{PH^3} = \frac{B_1/H^3}{3B_1/H^3 + S_2/H} \equiv \frac{1}{3 + \alpha_{12}^2} \tag{6.55}$$

The corresponding exact tip deflection is

$$\frac{B_1 w_{P/exact/E-S}(H)}{PH^3} = \frac{\alpha_{12} - \tanh\alpha_{12}}{\alpha_{12}^3} \tag{6.56}$$

TABLE 6.4

Exact Values and Estimates of the Dimensionless Tip
Deflection of an Euler Beam (B_1) Coupled to a Shear Beam
(S_2) Subject to a Uniform Load q_0 for Typical Material-
Geometry Parameters

$\alpha_{12} = \sqrt{\dfrac{S_2 L^2}{B_1}}$	$\dfrac{B_1 w_{q0/exact/E-S}(L)}{q_0 L^4}$	$\dfrac{B_1 w_{q0/estimate/E-S}(L)}{q_0 L^4}$	Error (%)
0.2	0.1231	0.1238	−0.60
0.6	0.1097	0.1147	−4.6
1.0	0.09035	0.10000	−10.7
2.0	0.05038	0.06250	−24.1
3.0	0.02982	0.03846	−29.0
4.0	0.01940	0.02500	−28.9
5.0	0.01358	0.01724	−27.0

As before, we see that Equations (6.55) and (6.56) differ markedly in form,
but the limiting cases of small and large α_{12} agree exactly. Further, our
numerical results in Table 6.5 indicate that Equations (6.55) and (6.56) produce remarkably similar results.

Coupled Timoshenko Beams: Castigliano Estimates (II)

We now illustrate the preceding extension of Castigliano's second theorem
by examining the displacements in another continuous coupled system,
a set of n simply supported Timoshenko beams, coupled in parallel and

TABLE 6.5

Exact Values and Estimates of the Dimensionless Tip Deflection of an Euler Beam
(B_1) Coupled to a Shear Beam (S_2) Subject to a Uniform Load P for Typical Material-
Geometry Parameters

$\alpha_{12} = \sqrt{\dfrac{S_2 L^2}{B_1}}$	$\dfrac{B_1 w_{P/exact/E-S}(L)}{PL^3}$	$\dfrac{B_1 w_{P/estimate/E-S}(L)}{PL^3}$	Error (%)
0.2	0.3281	0.3290	−3.0
0.6	0.2914	0.2976	−2.1
1.0	0.2384	0.2500	−4.9
2.0	0.1295	0.1429	−10.3
3.0	0.07426	0.08333	−12.2
4.0	0.04689	0.05263	−12.2
5.0	0.03200	0.03571	−11.6

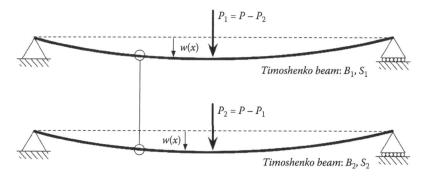

FIGURE 6.6
A system of two simply supported Timoshenko beams, each with bending stiffness B_i coupled in parallel to the shear stiffness S_i.

loaded by a uniform load q_0 (Figure 6.6). To say that these n-beams are coupled in parallel is to say, again, that each beam has the same total (bending plus shear) deflection. In that case, how are the loads divided among the n-beams?

The total shear and moment resultants to be supported by an n-beam system subject to the total uniform load q_0 and a concentrated (phantom) load P at the beam midpoint are

$$V(x) = \frac{P + q_0 L}{2} - q_0 x \qquad 0 \le x \le L/2$$

$$M(x) = \frac{P + q_0 L}{2} x - \frac{q_0 x^2}{2} \qquad 0 \le x \le L/2$$

(6.57)

Since the beams are symmetrically loaded, the shear $V(x)$ and moment $M(x)$ resultants are, respectively, antisymmetric and symmetric about the beam midpoint. We invoke the symmetry of the loading and write the complementary energy as

$$U^*_{n\text{-}Timo} = \sum_{i=1}^{n} \left[U^*_{si} + U^*_{bi} \right] = 2 \sum_{i=1}^{n} \left[\int_0^{L/2} \frac{(V_i(x))^2}{2 S_i} dx + \int_0^{L/2} \frac{(M_i(x))^2}{2 B_i} dx \right] \quad (6.58)$$

We now apportion both the distributed load $q_0 = q_{01} + q_{02} + ... + q_{0n}$ and the central load $P = P_1 + P_2 + ... + P_n$ just as we did in Equation (6.24) and we follow the same procedure we developed in Equations (6.23)–(6.36).

As part of that, we calculate the weighted resultant terms, which here turn out to be

$$\bar{V} = \frac{q_0 L^2}{8} \quad \text{and} \quad \bar{M} = \frac{5q_0 L^4}{384} \tag{6.59}$$

The dimensionless parameter corresponding to the weighted resultants (Equation 6.59) is

$$a^2 = \frac{48}{5} \tag{6.60}$$

Thus, for a parallel-coupled pair of simply supported Timoshenko beams supporting a uniform load, we can see that the midpoint deflection is

$$\frac{8S_1 w_0(0)}{q_0 L^2} = \frac{1 + 5\lambda_1^2/48}{1 + \dfrac{1 + 5\lambda_1^2/48}{1 + 5\lambda_2^2/48}\left(\dfrac{S_2}{S_1}\right)} = \frac{1 + 5\lambda_2^2/48}{1 + \dfrac{1 + 5\lambda_2^2/48}{1 + 5\lambda_1^2/48}\left(\dfrac{S_1}{S_2}\right)}\left(\frac{S_1}{S_2}\right) \tag{6.61}$$

We can also write Equation (6.61) in a form where $\bar{M}$ is the driver, which is why B_1 appears in the dimensionless amplitude (rather than S_1, as before):

$$\frac{384 B_1 w_0(0)}{5 q_0 L^4} = \frac{1 + 48/5\lambda_1^2}{1 + \dfrac{1 + 5\lambda_1^2/48}{1 + 5\lambda_2^2/48}\left(\dfrac{S_2}{S_1}\right)} \tag{6.62}$$

In Tables 6.6 and 6.7 we show values of the estimated midpoint deflection (Equation 6.62) alongside the corresponding exact results (which we do not

TABLE 6.6

Exact Values and Estimates of the Midpoint Deflection of a Pair of Coupled, Simply Supported, Timoshenko Beams Subject to a Uniform Load for Typical Material-Geometry Parameters ($\lambda_1^2 = 100.0$, $\lambda_2^2 = 1.0$)

B_2/B_1	$\alpha_{12} = \lambda_2\sqrt{\dfrac{B_2}{B_1}}$	$\dfrac{Bw_{0/exact}(0)}{q_0 L^4}$	$\dfrac{Bw_{0/estimate}(0)}{q_0 L^4}$	Error (%)
1.0	1.0	0.02591	0.02587	0.15
2.0	1.41	0.03558	0.03548	0.28
3.0	1.73	0.04374	0.04357	0.39

TABLE 6.7

Exact Values and Estimates of the Midpoint Deflection of a Pair
of Coupled, Simply Supported, Timoshenko Beams Subject to a
Concentrated Midpoint Load for Typical Material-Geometry
Parameters ($\lambda_1^2 = 100.0$, $\lambda_2^2 = 1.0$)

B_2/B_1	$\alpha_{12} = \lambda_2\sqrt{\dfrac{B_2}{B_1}}q$	$\dfrac{Bw_{P/exact}(0)}{PL^3}$	$\dfrac{Bw_{P/estimate}(0)}{PL^3}$	Error (%)
1.0	1.0	0.04250	0.04297	−1.11
2.0	1.41	0.05854	0.05971	−2.00
3.0	1.73	0.07218	0.07417	−2.76

provide here, although they can be obtained from the general solution in
Appendix A) for two simple Timoshenko beams. The agreement speaks for
itself—almost too good to be true!

Remarks on Modeling Buildings as Coupled-Beam Systems

It is worth asking how we might use a coupled Timoshenko beam to model
an actual structure, which to date it apparently has not. On the other hand,
the coupled Euler-shear beam combination, which is a subset of the coupled
pair of Timoshenko beams, has been used to model the dynamic response of
tall buildings, as we will discuss in Chapter 7. In either model, the question
would be the same: How do we connect the parameters of our coupled beam
models (e.g., the B_i and S_i in Equations 6.8, 6.9a, and 6.9b) to the properties of
an actual building?

Consider a building whose cross section comprises an external tube that
is connected to an internal core, as in braced-frame and shear wall-frame
buildings. We will take the external tube as a square that has sides of length
l with N_c columns arrayed on each side, each with an area A_{col} and modulus
E_{col}. We model the bending stiffness of the external tube as we would the
bending stiffness of the beam, which means we calculate the beam's second moment of area I_b in terms of the distribution of the areas of the columns arrayed around the perimeter of the tube. So, for bending around the
z-axis, for example, we treat the two sides normal to the transverse deflection
as flanges, and the second moment of the area of the columns along both
flanges is $I_f = N_{col}A_{col}l^2/2$.

The remaining two sides are treated as webs whose area consists of the
total area of the columns distributed over web height's l and corresponding web thickness $t_w = N_{col}A_{col}/l$. The second moment of each web area is
easily calculated as $I_w = N_{col}A_{col}l^2/12$, so that the total second area moment

for both webs is $I_w = N_{col}A_{col}l^2/6$. Note that this elementary distribution model assumes that the webs contribute substantially to the bending stiffness of the tube. Note as well that we can also calculate the second moment of the webs by noting that the columns on the web are spaced at uniform intervals of $\Delta = l/N_{col}$. Then we would calculate their resulting second moment by summing series much as we did for beams and trusses in Chapter 3; that is,

$$I_w = 2A_{col}\sum_{i=1}^{N_{col}/2}(i\Delta)^2 = 2A_{col}\Delta^2\left(\frac{N_{col}/2}{6}\right)\left(\frac{N_{col}}{2}+1\right)\left(\frac{N_{col}}{2}+1\right) \cong N_{col}A_{col}l^2/12 \quad (6.63)$$

With either calculation of I_w, we find the bending stiffness of the tube to be

$$B = E_{col}(I_f + 2I_w) = 2E_{col}N_{col}A_{col}l^2/3 \quad (6.64)$$

Note that our model of the tube bending stiffness (Equation 6.64) assumes that any diagonal elements on the building's perimeter do not contribute to the bending stiffness. Also, we should remember that since Equation (6.64) is based on elementary Euler–Bernoulli beam theory, it allows the curvature of the tube (and building) to vary along the building's length (or height).

To model the shear stiffness, we need to decide—or make assumptions about—how the shear is allocated among the structure's elements. In the same way that we essentially calculated the bending stiffness B in terms of an equivalent EI, we might proceed analogously to calculate the shear stiffness S in terms of GA. One model of the stiffness of a separate shear-carrying core would view the essentially solid core as a simple shear beam. Still further, we might argue that the shear is carried by a combination of the cross-sectional area A_{core} of such a solid core, augmented by the net cross-sectional areas of the columns—that is,

$$S = G_{core}A_{core} + 4G_{col}N_{col}A_{col} \quad (6.65)$$

Still further, if we supposed that the external frame included diagonals, as we often see in modern skyscrapers, they would contribute to shear stiffness. In such *diagrid* systems, the diagonals are assumed to be purely axial, truss-like elements: They thus carry the shear load by their own extension (or compression). So with a pair of diagonals on each side of length d and connecting floors separated by height h, we can see that each diagonal is inclined to the horizontal by the angle $\theta = \sin^{-1}(h/d)$. Then we would augment the shear stiffness given in Equation (6.65) by accounting for an added shear force $V = 4F\cos\theta$ carried by the tension and compression F and $-F$ in each diagonal pair.

These diagonals support a shear strain $\gamma = \Delta/h$, where Δ is the relative transverse displacement of the building between any two floors. That shear strain translates into an extension (and compression) of each diagonal, $\varepsilon_{diag} = (\Delta/h)\sin\theta\cos\theta$, which means that the magnitude of F is simply $F = E_{diag}A_{diag}\varepsilon_{diag}$. For such a diagrid system, it is then appropriate to augment the shear stiffness of a diagonal-braced frame with a term attributed to those diagonals,

$$S_{diag} = 4\sin\theta\cos^2\theta E_{col}A_{diag} \qquad (6.66)$$

so that the total shear stiffness of a diagrid building, accounting for the tube and the diagonal frame, would be

$$S_{tube+diag} = G_{core}A_{core} + 4G_{col}N_{col}A_{col} + 4\sin\theta\cos^2\theta E_{col}A_{diag} \qquad (6.67)$$

We can then combine Equation (6.64) with Equation (6.65) or (6.67)—or perhaps some altogether different model of the shear distribution—to estimate the static stiffness of a dual-system building. The static stiffness can be easily extended to very tall buildings that might be modeled as serially connected modules of individual height L. Further, we can use such static stiffness expressions to derive quick gross stiffness (and then weight) estimates and even to perform preliminary optimization studies. Still further, having noted the different weights or emphases accorded to the bending and shear stiffnesses for the three loadings considered here, we could imagine a variety of studies in which we can readily estimate (and optimize) the static stiffness for a variety of loading conditions.

Conclusions

We devoted this chapter to providing simple estimators of the deflections of systems of Timoshenko beams coupled in parallel. We derived those estimators by applying a recently developed extension of Castigliano's theorems for coupled systems and we validated them by comparing the resulting deflection estimates to exact solutions for the static response of systems of coupled Timoshenko beams. The estimators we derived using the Castigliano-based approach are clearly simpler and easier to apply, which could be very useful for rapid preliminary design assessments. The deflection estimates they produced agreed quite well with the exact results obtained with a formal solution for a coupled system: The limiting cases agreed exactly and the general numerical results were generally quite close to the exact results. Thus, the estimators can be used easily and with confidence.

We also saw that the effective stiffness of the system of two coupled beams depends on four material properties (E_1, G_1, E_2, G_2) and six geometric properties (k_1, I_1, A_1, k_2, I_2, A_2) that were encapsulated in just two dimensionless parameters, λ_1^2 and λ_2^2. We noted that while the contributions of the bending and shear stiffnesses are weighted differently for different load distributions, the weighting factor a^2 does not vary much among the four cases examined previously, as we saw in Table 6.1. In fact, the results in Table 6.1 suggest that it may not be worth distinguishing among load types for preliminary estimates of the deflections of the beam systems.

Finally, it is worth noting that the approach we derived here makes it rather easy to extend the nature of the models used to model the tip deflections or drifts of tall buildings by extending the model based on coupling an Euler beam to a shear beam with an extended model in which bending and shear are supported in each of the two coupled Timoshenko beams. This enables us, for example, to incorporate the shear flexibility of the tubes and the bending flexibility of the shear cores in a tube-and-core system.

Appendix A: Exact Solution for Coupled Timoshenko Beams

For a pair of Timoshenko beams subject to a transverse load, $q(x) = q_0 f(x)$, we introduce the dimensionless variables

$$(W, \Psi_i) = \frac{B}{q_0 L^3} \left(\frac{w}{L}, \psi_i \right) \qquad (i = 1, 2) \tag{6.68}$$

and a dimensionless coordinate, $\xi = x/L$, in which case our governing Equation (6.5) becomes

$$\Psi_i''(\xi) + \lambda_i^2 (W'(\xi) - \Psi_i) = 0 \qquad (i = 1, 2) \tag{6.69a}$$

$$W''(\xi) - \frac{S_1}{S} \Psi_1'(\xi) - \frac{S_2}{S} \Psi_2'(\xi) = -f(\xi)/\lambda^2 \tag{6.69b}$$

As is often the case with complex systems of linear differential equations, multiple paths might be taken to solve them—although those paths must always produce the same result! Our approach will be to cast Equations

(6.69a) and (6.69b) in matrix form after introducing the derivative operator $D = d/d\xi$ so that our equilibrium equations are now

$$
\begin{bmatrix}
D^2 - \lambda_1^2 & 0 & \lambda_1^2 D \\
0 & D^2 - \lambda_2^2 & \lambda_2^2 D \\
(S_1/S)D & (S_2/S)D & -D^2
\end{bmatrix}
\begin{Bmatrix}
\Psi_1 \\ \Psi_2 \\ W
\end{Bmatrix}
=
\begin{Bmatrix}
0 \\ 0 \\ f(\xi)/\lambda^2
\end{Bmatrix}
\tag{6.70}
$$

We can now use Cramer's rule to solve Equation (6.70) to find the individual uncoupled differential equations, with the net result being that

$$
\lambda^2 D^4 (D^2 - \gamma^2)\Psi_1 = \lambda_1^2 D(D^2 - \lambda_2^2)f(\xi)
$$
$$
\lambda^2 D^4 (D^2 - \gamma^2)\Psi_2 = \lambda_2^2 D(D^2 - \lambda_1^2)f(\xi)
\tag{6.71}
$$
$$
\lambda^2 D^4 (D^2 - \gamma^2)W = -(D^2 - \lambda_1^2)(D^2 - \lambda_2^2)f(\xi)
$$

We see in Equation (6.71) that the same differential operators act on each of the three displacement variables. We also see that for a uniform load the two bending rotations are determined by their homogeneous solutions because $Df(\xi) = D(1) = 0$. We further see that for a uniform load, the transverse displacement is determined from

$$
\lambda^2 D^4 (D^2 - \gamma^2)W = -\lambda_1^2 \lambda_2^2
\tag{6.72}
$$

Since all of our differential equations have constant coefficients, the displacement terms are proportional to $e^{\alpha\xi}$, so it is readily seen from the homogeneous versions of either Equation (6.71) or (6.72) that the characteristic roots α are

$$
\alpha = 0, 0, 0, 0, \gamma, -\gamma
\tag{6.73}
$$

Given the four-times repeated root of 0 and the roots $\pm\gamma$, it is clear that the homogeneous solution of Equation (6.72) is

$$
w_h(x) = C_0 + C_1\left(\frac{x}{L}\right) + C_2\left(\frac{x}{L}\right)^2 + C_3\left(\frac{x}{L}\right)^3 + C_4\cosh\frac{\gamma x}{L} + C_5\sinh\frac{\gamma x}{L}
\tag{6.74}
$$

Clearly, our solution (Equation 6.7a) reflects the homogeneous solution (Equation 6.74).

For a uniform load, the particular solution to Equation (6.71) is easily found to be $\xi^4/24$. The six constants of integration (C_0, C_1, C_2, C_3, C_4, C_5) are then determined by satisfying the appropriate boundary conditions. For a cantilever system, the correct boundary conditions are given by Equations (6.10a) and (6.10b), and we then find the constants are

$$
C_0 = \frac{\left(\lambda_1^2 - \gamma^2\right)\left(\lambda_2^2 - \gamma^2\right)}{\gamma^4 \lambda_1^2 \lambda_2^2}\left(\frac{1 + \gamma \sinh \gamma}{\cosh \gamma}\right)
$$

$$
C_1 = \frac{-1}{\gamma^2} + \frac{\lambda_1^2 + \lambda_2^2}{\lambda_1^2 \lambda_2^2}
$$

$$
C_2 = \frac{1}{2\gamma^2} + \frac{1}{4} - \frac{\lambda_1^2 + \lambda_2^2}{2\lambda_1^2 \lambda_2^2} \tag{6.75}
$$

$$
C_3 = -1/6
$$

$$
C_4 = -\frac{\left(\lambda_1^2 - \gamma^2\right)\left(\lambda_2^2 - \gamma^2\right)}{\gamma^4 \lambda_1^2 \lambda_2^2}\left(\frac{1 + \gamma \sinh \gamma}{\cosh \gamma}\right)
$$

$$
C_5 = \frac{\left(\lambda_1^2 - \gamma^2\right)\left(\lambda_2^2 - \gamma^2\right)}{\gamma^3 \lambda_1^2 \lambda_2^2}
$$

As a final note, it can be seen from the form of the governing equations and the boundary conditions that the indices 1, 2 can be interchanged in the expressions for the Euler angles and the displacement. Thus, the solution is symmetric with regard to the order of beams 1 and 2, as is also evident from the structure of the governing Equation (6.69).

Appendix B: The Coupled Euler–Shear System as a Limit

We now show the analytical details of how the deflected shape of a uniformly loaded cantilever system (Equation 6.11) produces the limit of the corresponding deflection of a cantilever system wherein an Euler beam is tied to a shear beam (Equation 6.15). We start first by recalling that if beam 1 is the Euler beam and beam 2 is the shear beam, this means that $S_1 \to \infty$ and $B_2 \to \infty$. It then readily follows from Equation (6.9a) that $\lambda_1^2 \to \infty$ and $\lambda_2^2 \to 0$, and Equation (6.9b) has a finite, nonzero limit of $\gamma^2 \to \alpha_{12}^2$. This last limit also means we can write $\lambda_1^2 = (S_1/S_2)\alpha_{12}^2 \to \infty$ and $\lambda_2^2 = (B_1/B_2)\alpha_{12}^2 \to 0$. We also note that the dimensional coefficient that defines Equation (6.11) can be rewritten as

$$
\frac{q_0 L^4}{B} = \frac{q_0 L^2}{S_2 \gamma^2}\left(\frac{\lambda_1^2 \lambda_2^2}{1 + S_1/S_2}\right) \tag{6.76}
$$

Now we can rewrite our exact solution (Equation 6.11) for the cantilever system as

$$w_{q0/exact}(L) = \frac{q_0 L^2}{S_2 \gamma^2} \left[\begin{array}{l} \left(\dfrac{1}{8} - \dfrac{1}{2\gamma^2}\right)\dfrac{\lambda_1^2 \lambda_2^2}{(1+S_1/S_2)} + \dfrac{\lambda_1^2 + \lambda_2^2}{2(1+S_1/S_2)} \\[4mm] - \dfrac{\left(\gamma^2 - \lambda_1^2\right)\left(\gamma^2 - \lambda_2^2\right)}{\gamma^4 (1+S_1/S_2)}\left(1 - \dfrac{1+\gamma \sinh \gamma}{\cosh \gamma}\right) \end{array} \right]$$ (6.77)

If we now examine each of the three terms in Equation (6.77), noting especially their common denominators, we see the following limits as $S_1 \rightarrow \infty$ and $B_2 \rightarrow \infty$:

$$\frac{\lambda_1^2 \lambda_2^2}{(1+S_1/S_2)} = \frac{\lambda_2^2 L^2 S_1}{(1+S_1/S_2)B_1} \rightarrow \alpha_{12}^2 \lambda_2^2 \rightarrow 0$$

$$\frac{\lambda_1^2 + \lambda_2^2}{2(1+S_1/S_2)} = \frac{\lambda_2^2 + S_1 L^2/B_1}{2(1+S_1/S_2)} \rightarrow \frac{S_1 L^2/B_1}{2 S_1/S_2} = \frac{\alpha_{12}^2}{2}$$ (6.78)

$$\frac{\left(\gamma^2 - \lambda_1^2\right)\left(\gamma^2 - \lambda_2^2\right)}{\gamma^4 (1+S_1/S_2)} \rightarrow \frac{(1 - B_1/B_2)(1 - S_1/S_2)}{(1+S_1/S_2)} \rightarrow -1$$

Now we insert the limits (Equation 6.78) into the deflection (Equation 6.77) to find the coupled Euler–shear limit that is exactly Equation (6.15):

$$w_{q0/exact/E-S}(L) = \frac{q_0 L^2}{\alpha_{12}^2 S_2 \cosh \alpha_{12}} \left[\left(1 + \frac{\alpha_{12}^2}{2}\right)\cosh \alpha_{12} - \left(1 + \alpha_{12} \sinh \alpha_{12}\right) \right]$$ (6.79)

Appendix C: Special Cases for Two Timoshenko Cantilevers

We now consider a coupled Timoshenko system under a load distributed linearly along its length: Thus we approximate a seismic input. The load apportionment ratio is found to be

$$r_{1/q_1} = \frac{\dfrac{B_1}{1+40/11\lambda_1^2}}{\dfrac{B_1}{1+40/11\lambda_1^2} + \dfrac{B_2}{1+40/11\lambda_2^2}} = \frac{1}{1 + \dfrac{\left(1+40/11\lambda_1^2\right)}{\left(1+40/11\lambda_2^2\right)}\left(\dfrac{B_2}{B_1}\right)}$$ (6.80)

and the corresponding tip deflection or drift estimate in this instance is

$$\frac{B_1 w_{q_1/estimate}(H)}{q_1 H^4} = \frac{(40/33)B_1}{\dfrac{B_1}{1+40/11\lambda_1^2} + \dfrac{B_2}{1+40/11\lambda_2^2}} \tag{6.81}$$

For the specific case of an Euler beam (beam 1; $S_1 \to \infty$, $\lambda_1^2 \to \infty$) coupled to a shear beam (beam 2; $B_2 \to \infty$, $\lambda_2^2 \to \infty$), the concomitant tip deflection follows from Equations (6.80) and (6.81) as

$$\frac{B_1 w_{q_1/estimate/E-S}(H)}{q_1 H^4} = \frac{B_1/3H^3}{40B_1/11H^3 + S_2/H} = \frac{1}{3\left(40/11 + \alpha_{12}^2\right)} \tag{6.82}$$

The exact counterpart in this instance can be found to be

$$\frac{B_1 w_{q_1/exact/E-S}(H)}{q_1 H^4} = \frac{3\left(2 - \alpha_{12}^2\right)\sinh\alpha_{12} + 2\alpha_{12}^3 \cosh\alpha_{12} - 6\alpha_{12}}{6\alpha_{12}^5 \cosh\alpha_{12}} \tag{6.83}$$

If we examine the case of a surface load that varies quadratically—in fact, as a parabola—along the length of a coupled pair of Timoshenko cantilevers, the load apportionment factor is found to be

$$r_{1/q_2} = \frac{1}{1 + \dfrac{1 + 13\lambda_1^2/45}{1 + 13\lambda_2^2/45}\left(\dfrac{S_2}{S_1}\right)} \tag{6.84}$$

and the tip deflection or drift estimate is

$$\frac{B_1 w_{q_2/estimate}(H)}{q_2 H^4} = \frac{(13/180)B_1}{\dfrac{B_1}{1+45/13\lambda_1^2} + \dfrac{B_2}{1+45/13\lambda_2^2}} \tag{6.85}$$

Once again specialized for the coupling of an Euler beam to a shear beam, Equation (6.85) yields a corresponding tip deflection

$$\frac{B_1 w_{q_2/estimate/E-S}(H)}{q_2 H^4} = \frac{B_1/4H^3}{45B_1/13H^3 + S_2/H} = \frac{1}{4(45/13 + \alpha_{12}^2)} \tag{6.86}$$

whose exact counterpart can be shown to be

$$\frac{B_1 w_{q_2/exact/E-S}(H)}{q_2 H^4} = \frac{3\left(8+\alpha_{12}^4\right)\cosh\alpha_{12} - 4\alpha_{12}^3\sinh\alpha_{12} - 12\left(2+\alpha_{12}^2\right)}{12\alpha_{12}^6\cosh\alpha_{12}} \quad (6.87)$$

The estimates of the tip deflection given by Equations (6.81), (6.82), (6.86), and (6.87) agree quite well with their more complex exact equivalents: The analytical results produce identical limiting values for both small and large values of α_{12}. Further, the numerical values, which we do not present here, agree within fractions of 1% at small and large values of α_{12} and differ by less than 10% at their worst, for values of $\alpha_{12} \approx O(1)$.

Finally, consider a coupled pair of Timoshenko beams loaded by a tip moment M^*. In this case the shear force $V(x) = 0$, so Equation (6.31) must be modified in the light of the definition (Equation 6.32) to read

$$w_{q(x)}(L) = \sum_{i=1}^{2}\left[\left(\frac{r_i^2}{S_i}\right)\left(\lambda_i^2\left(\frac{\bar{M}}{L^2}\right)\right)\right] \equiv \sum_{i=1}^{2}\left[\left(\frac{r_i^2}{B_i}\right)\bar{M}\right] \quad (6.88)$$

In this instance $\bar{M} = \int_0^L M^*(x-L)dx = M^*L^2/2$, so the tip deflection of the beam loaded by the tip moment follows from Equation (6.87) as

$$w_{M^*}(L) = \sum_{i=1}^{2}\left[\left(\frac{r_i^2 L^2}{2B_i}\right)M^*\right] \quad (6.89)$$

Equation (6.89) shows us that the stiffness of each beam in response to the tip moment M^* can be identified as $K_{TM^*i} = 2B_i/L^2$, with a corresponding apportionment factor $r = K_{TM^*1}/(K_{TM^*1} + K_{TM^*2})$, which in turn allows us to write the tip displacement as

$$w_{M^*}(L) = \frac{M^*}{K_{TM^*1} + K_{TM^*2}} = \frac{M^*L^2}{2(B_1 + B_2)} \quad (6.90)$$

It is important that we recognize that the two preceding results apportion the loads on the coupled Timoshenko beam model in accord with satisfying equilibrium and ensuring compatible displacements, which is just the outcome we get by applying Castigliano's second theorem. It happens that the usual discrete modeling reflected previously yields the same result for the discrete tip load and moment.

Bibliography

Dym, C. L. 1997. *Structural modeling and analysis.* New York: Cambridge University Press.
———. 2010. Extending Castigliano's theorems to model the behavior of coupled systems. *Journal of Applied Mechanics* 77 (6): 061005–1-6.
Dym, C. L., and I. H. Shames. 1973. *Variational methods in solid mechanics.* New York: McGraw–Hill.
Englekirk, R. E. 1994. *Steel structures: Controlling behavior through design.* New York: John Wiley & Sons.
Moon, K., J. J. Connor, and J. E. Fernandez. 2007. Diagrid structural systems for tall buildings: Characteristics and methodology for preliminary design. *Structural Design of Tall and Special Buildings* 16 (2): 205–230.
Stafford Smith, B., and A. Coull. 1991. *Tall building structures: Analysis and design.* New York: John Wiley & Sons.
Taranath, B. S. 1988. *Structural analysis and design of tall buildings.* New York: McGraw-Hill.

Problems

6.1 For the critical case $\lambda_1^2 = \lambda_2^2$ (i.e., $S_1/S_2 = B_1/B_2$), show that $\lambda^2 = \lambda_1^2 = \lambda_2^2$ and $\gamma_{crit}^2 = \lambda^2$.

6.2 For the critical case $\lambda_1^2 = \lambda_2^2$, find the explicit solutions for the two Euler angles $\psi_1(\xi)$, $\psi_2(\xi)$ for a cantilever supporting a uniform load q_0. Verify the constants of integration given in Equation (6.70).

6.3 Determine the exact transverse displacement at the center ($\xi = 0$) of a system of two simply supported Timoshenko beams subject to a uniform load q_0 for the critical case $\lambda_1^2 = \lambda_2^2$. Assume that each beam satisfies the conditions $M_i(\pm 1/2) = 0$.

6.4 Determine the exact transverse displacement at the center ($\xi = 0$) of a system of two simply supported Timoshenko beams subject to a centrally applied concentrated load P for the critical case $\lambda_1^2 = \lambda_2^2$. Assume that each beam satisfies the conditions $M_i(\pm 1/2) = 0$.

6.5 Determine the exact transverse displacement at the center ($\xi = 0$) of a simply supported system of an Euler beam coupled to a shear beam and subject to a uniform load q_0.

6.6 Determine the exact transverse displacement at the center ($\xi = 0$) of a simply supported system of an Euler beam coupled to a shear beam and subject to a centrally applied concentrated load P.

6.7 Using Castigliano's second theorem extended, find an approximate solution for the transverse displacement at the center ($\xi = 0$) of a system of two simply supported Timoshenko beams subject to a uniform load q_0 for the critical case $\lambda_1^2 = \lambda_2^2$. Assume that each beam satisfies the conditions $M_i(\pm 1/2) = 0$.

6.8   Using Castigliano's second theorem extended, find an approxi-
       mate solution for the transverse displacement at the center ($\xi = 0$)
       of a system of two simply supported Timoshenko beams subject to
       a centrally applied concentrated load P for the critical case $\lambda_1^2 = \lambda_2^2$.
       Assume that each beam satisfies the conditions $M_i(\pm 1/2) = 0$.

6.9   Using Castigliano's second theorem extended, find an approximate
       solution for the transverse displacement at the center ($\xi = 0$) of a sim-
       ply supported system of an Euler beam coupled to a shear beam and
       subject to a uniform load q_0.

6.10   Using Castigliano's second theorem extended, find an approximate
       solution for the transverse displacement at the center ($\xi = 0$) of a sim-
       ply supported system of an Euler beam coupled to a shear beam and
       subject to a centrally applied concentrated load P.

7

Buildings Modeled as Coupled Beams: Natural Frequency Estimates

Summary

If we're going to design structural systems to withstand earthquakes, wind, or other dynamic forces, we have to be able to calculate or estimate the natural frequencies of such systems as a central measure of their dynamic properties. Therefore, as engineers, we want to be able to estimate these fundamental frequencies as quickly and efficiently as possible. While modern computational techniques enable us to calculate such frequencies for very complex configurations, geometries, and loads, we always have reasons to do quick, meaningful "back of the envelope" calculations that confirm (or deny) our intuition, as well as provide a framework for interpreting and confirming complex computer modeling.

   Our aim in this chapter is to present the Rayleigh quotient as a powerful (and beautiful) tool for developing such estimates. From a modeling perspective, we will first show how the basic statement of Rayleigh's quotient, properly rendered dimensionless, can be used to estimate the natural frequencies of Euler beams. Then, and once again in the context of using coupled systems of beams to model the behavior of an entire building, we show how natural frequencies of such models depend on the beam's length (or the building's height) to conform with empirical data. We then show that the Rayleigh quotient estimates compare rather well with some exact analytical results.

Introduction

Modeling and estimating the behavior of tall buildings is another area where simple "back of the envelope" models are valued in engineering practice. Indeed, there is a substantial body of work in which researchers have tried to build models of buildings in terms of beam components.

As we noted in Chapter 6, the static models of analysis assume series models of behavior wherein bending and shear deflections of beam components (or building drifts) add to form the total deflection or drift. We now explore the same issues in the "dynamics" context of estimating the fundamental frequencies of such buildings.

A specific concern in the development of estimates of the natural frequency of tall buildings has been the proper characterization of the dependence of frequency on the building height L. Empirical data and practical experience have long suggested that, for shear wall-frame structures, the natural frequency varies as $1/L$, while the fundamental frequencies of the two most familiar beam models, the Euler–Bernoulli cantilever and the Timoshenko cantilever, vary as $1/L^2$. However, when we couple an Euler beam in parallel to a shear beam by requiring that both undergo the same deflection, we find that the fundamental frequency varies as $1/L$ for realistic conditions. We identify such realistic conditions in terms of a parameter α that is the ratio of the coupled system's (or building's) *shear stiffness, S/L,* to its *bending stiffness, B/L^3*—namely, $\alpha^2 = SL^2/B$. The ranges of interest of values of the building parameter α include the following:

- $0 \le \alpha \le 1.5$ for shear wall and braced frame buildings
- $1.5 \le \alpha \le 5$ for *dual-system* buildings combining moment-resisting frames with shear walls or braced frames
- $5 \le \alpha \le 20$ for moment-resisting frames

Shear wall-frame or tube-and-core buildings fall into the middle category that combines moment-resisting tubes with a core or frame to carry the shear. The coupled Euler-and-shear beam model shows a frequency dependence of $1/L$ for larger values of α^2, making it appropriate for modeling shear wall-frame (e.g., tube and core) construction. And as we shall see (and as has been demonstrated elsewhere), the coupled Euler-and-shear beam model has been used successfully to model the dynamic response of a building comprising a shear core and a bent tube.

Rayleigh Quotients for Elementary Beams

We start by building on our introduction in Chapter 5 of Rayleigh quotients for discrete systems by showing how well they work to estimate the fundamental frequencies of an elementary cantilever. The equations of motion of a continuous system can be determined by minimizing its Lagrangian L. We construct the Lagrangian of a simple cantilever with length L and constant bending stiffness B, mass per unit volume ρ, and area A as the difference between the kinetic energy terms and the total potential energy for that elementary beam. Thus, with

$$\mathcal{L} = T - (U + V) \tag{7.1}$$

the kinetic energy for the cantilever is

$$T = \frac{\rho A}{2} \int_0^L \left(\frac{\partial w(x,t)}{\partial t} \right)^2 dx \tag{7.2}$$

and the strain energy for this Euler–Bernoulli beam is

$$U = \frac{B}{2} \int_0^L \left(\frac{\partial^2 w(x,t)}{\partial x^2} \right)^2 dx \tag{7.3}$$

We will define the total potential of the applied loading, V, later (Equation 7.20). In view of Equations (7.1)–(7.3), we now take the Lagrangian for an Euler beam as

$$\mathcal{L} = \frac{\rho A}{2} \int_0^L \left(\frac{\partial w(x,t)}{\partial t} \right)^2 dx - \frac{B}{2} \int_0^L \left(\frac{\partial^2 w(x,t)}{\partial x^2} \right)^2 dx \tag{7.4}$$

We apply Hamilton's principle to this Lagrangian to derive the equation of motion for our Euler beam: We write $\delta L = 0$ and perform the usual integrations by parts to find:

$$\delta\mathcal{L} = 0 = -B \int_0^L \left[\rho A \frac{\partial^2 w(x,t)}{\partial t^2} + B \frac{\partial^4 w(x,t)}{\partial x^4} \right] \delta w \, dx$$

$$+ \left[B \frac{\partial^2 w(x,t)}{\partial x^2} \delta \left(\frac{\partial w}{\partial x} \right) - B \frac{\partial^3 w(x,t)}{\partial x^3} \delta w \right]_0^L \tag{7.5}$$

Equation (7.5) contains the standard equation of motion for a vibrating beam:

$$B \frac{\partial^4 w(x,t)}{\partial x^4} + \rho A \frac{\partial^2 w(x,t)}{\partial t^2} = 0 \tag{7.6}$$

along with the duals of the usual boundary condition choices for Euler–Bernoulli beams. We find the beam's fundamental frequency by assuming harmonic motion in time:

$$w(x,t) = W(x)\cos\omega t \tag{7.7}$$

Then Equation (7.6) becomes

$$B\frac{d^4 W(x)}{dx^4} - \rho A\omega^2 W(x) = 0 \tag{7.8}$$

Equation (7.8) is a classical eigenvalue problem whose solution is, for a simple cantilever supported at $x = 0$,

$$W(x) = W_0\left[\cosh k_1 x - \cos k_1 x - \frac{\cosh k_1 L + \cos k_1 L}{\sinh k_1 L + \sin k_1 L}(\sinh k_1 x - \sin k_1 x)\right] \tag{7.9}$$

where $k^4 \triangleq eA\omega^2/B$ and the mode shape (Equation 7.9) corresponds to the fundamental or lowest frequency, with the value $k_1 L = 1.875$ found as the root of the transcendental equation

$$\cosh k_1 L \cos k_1 L + 1 = 0. \tag{7.10}$$

The corresponding fundamental frequency is, then,

$$\omega_{cant} = (1.875)^2\sqrt{\frac{B}{\rho A L^4}} = 3.52\sqrt{\frac{B}{\rho A L^4}} \tag{7.11}$$

Equations (7.7)–(7.11) comprise a formal exact solution for the cantilever's natural frequency. We see that even for this simple problem, the mode shape (Equation 7.9) is a complicated function and the frequency (Equation 7.11) is itself only determined after we solve the transcendental Equation (7.10). So, we can easily see why Rayleigh and others of his time were looking for more manageable approximations. Remember that this was long before the age of computers and that, nowadays, even with modern computational power, engineers continue to seek simpler approximations or estimates.

To get such frequency estimates, we use the precursors of the Lagrangian (Equation 7.4) to construct a Rayleigh quotient for the beam, just as we did for discrete systems in Chapter 5. We set the maximum (in time) potential and kinetic energies equal one to another, from which we find that the frequency of this harmonic motion is given by the quotient

$$\omega^2 = \left(\frac{B}{\rho A L^4}\right) \frac{\displaystyle\int_0^1 (W''(\xi))^2 \, d\xi}{\displaystyle\int_0^1 (W(\xi))^2 \, d\xi} \tag{7.12}$$

Thus, if we substitute any suitably chosen function $W(\xi) = W(x/L)$ into the two integrals in the quotient (Equation 7.12), we immediately generate an estimate of the natural frequency ω. We now illustrate the beauty and power of the Rayleigh quotient (Equation 7.12) by showing how easily we can get a very accurate approximation of that frequency.

We know that a cantilever must have zero displacement and slope at its sole support, say, $x = 0$, and that the free end must have zero bending and shear resultants. In the spirit of approximating the fundamental mode shape, let us assume that

$$W_1(x) = W_{01}\left(1 - \cos\frac{\pi x}{2L}\right) \tag{7.13}$$

If we substitute the assumed mode shape (Equation 7.13) into the Rayleigh quotient (Equation 7.12), we get the following result:

$$\omega_1 = 3.67 \sqrt{\frac{B}{\rho A L^4}} = 1.04\omega_{cant} \tag{7.14}$$

So, with a simple, intuitively pleasing mode shape, we get an approximation that is within 4% of the exact answer! Further, let us take the deflected shape of a cantilever under a uniform static load as our mode shape; that is, let

$$W_2(x) = W_{02}\left[\left(\frac{x}{L}\right)^4 - 4\left(\frac{x}{L}\right)^3 + 6\left(\frac{x}{L}\right)^2\right] \tag{7.15}$$

Here, when we substitute the assumed mode shape (Equation 7.15) into the Rayleigh quotient (Equation 7.12), we find that

$$\omega_2 = 3.53 \sqrt{\frac{B}{\rho A L^4}} = 1.00\omega_{cant} \tag{7.16}$$

We once again get an essentially exact approximation; in fact, $\omega_2 = 1.0028\omega_{cant}$ here, but Equation (7.16) is close enough for engineering practice! And lastly, suppose we assume an even simpler form with no obvious physical connection except that it and its slope are zero at the support:

$$W_3(x) = W_{03}\left(\frac{x}{L}\right)^2 \tag{7.17}$$

When we substitute this mode shape (Equation 7.17) into the Rayleigh quotient (Equation 7.12), we find that

$$\omega_3 = 4.47\sqrt{\frac{B}{\rho A L^4}} = 1.27\omega_{cant} \tag{7.18}$$

So, even with an "out of the blue" mode shape, we get within 27%, which might be a reasonable estimate when we want a really quick answer at virtually no arithmetic cost. The most important conclusion we should draw from these three examples is that we can readily obtain quite good estimates of fundamental frequency using the Rayleigh quotient, even when our mode shape guesses are less than perfect.

Finally, it is worth noting that the Rayleigh quotient (Equation 7.12) is easily extended to account for other effects and other geometries. For example, suppose we wanted to incorporate the vertical load due to a vertically oriented beam's own weight per unit length, q_0. The effect of that load can be developed in terms of the moment it produces in the beam at a location x from the fixed support at the base:

$$M_{q_0}(x,t) = -\int_x^L q_0[w(\varsigma,t) - w(x,t)]d\varsigma \tag{7.19}$$

For cantilever boundary conditions, this result can also be expressed as an equivalent potential of the axial load q_0:

$$V_{q_0} = -\frac{q_0 L}{2}\int_0^L \left(1 - \frac{x}{L}\right)\left(\frac{\partial w(x,t)}{\partial x}\right)^2 dx \tag{7.20}$$

The potential energy term (Equation 7.20) produces the second-order effect of the gravity loads, known to structural engineers as the P–Δ effect. In fact, we find the corresponding Rayleigh quotient including the self-weight by substituting the potential energy into the energy balance equation and thus see that

$$\omega^2 = \frac{\left(\dfrac{B}{\rho A L^4}\right) \displaystyle\int_0^1 (W''(\xi))^2 \, d\xi - \left(\dfrac{g}{L}\right) \displaystyle\int_0^1 (W'(\xi))^2 \, d\xi}{\displaystyle\int_0^1 (W(\xi))^2 \, d\xi} \tag{7.21}$$

It turns out, though, that the factor (g/L) is generally sufficiently small compared to $(B/\rho A L^4)$ that we can neglect the beam's self-weight when we calculate the natural frequency. That is, as we noted before, the P–Δ effect is a second-order effect.

However, it is worth noting that the form of Equation (7.20) is also identical in form—save for a different load—to the potential energy of an axial load P applied along the axis of an elastic column. In that case we would replace $(q_0 L)$ with P, which effectively assumes that P is positive in compression. (Notice here that we are transforming our elementary beam into an axially loaded column!) Then we can readily show that the Rayleigh quotient for the free vibration of this axially loaded column is

$$\omega^2 = \left(\frac{B}{\rho A L^4}\right) \frac{\displaystyle\int_0^1 (W''(\xi))^2 \, d\xi - \left(\dfrac{P}{P_E}\right) \displaystyle\int_0^1 (W'(\xi))^2 \, d\xi}{\displaystyle\int_0^1 (W(\xi))^2 \, d\xi} \tag{7.22}$$

We have introduced the Euler buckling load of a simply supported column, $P_E = \pi^2 B/L^2$, into Equation (7.22) to maintain the dimensionless formulation of our Rayleigh quotient. What is really interesting about the quotient (Equation 7.22) is that the arithmetic sign of its numerator could change with the value of P. Thus, were P to be negative, as if we were shortening the beam or column's axis, the frequency would increase, reflecting the increase in the lateral stiffness that results from the tension induced by $P < 0$. On the other hand, for $P > 0$, the frequency gets smaller. In fact, we see that the frequency vanishes whenever the axial load takes on the value

$$P = P_E \frac{\displaystyle\int_0^1 (W''(\xi))^2 \, d\xi}{\displaystyle\int_0^1 (W'(\xi))^2 \, d\xi} \tag{7.23}$$

So now we have found another Rayleigh quotient; this one enables us to estimate values of the buckling loads of axially loaded columns because the quotient (Equation 7.23) defines the value at which the column's natural frequency passes through zero as it transitions from stability characterized by harmonic vibration (see Equation 7.8) to instability characterized by vibration that grows exponentially (see Equation 7.5 when $\omega^2 < 0$).

Beams and Models of Buildings

As we noted in Chapter 6, engineers and researchers have long been trying to model overall building behavior in terms of the response of beams or collections of beams. For example, it is tempting to think of a tall, slender building as a vertically oriented cantilever beam. Unfortunately, this notion is undermined by a simple, yet ugly, fact: Empirical measurements of the natural frequency of tall buildings show that these frequencies vary inversely with their height, which is not the dependency shown by the familiar models of beam behavior. So, we want to talk about that for a bit.

We have already calculated the natural frequency of an elementary cantilever beam, and we restate the result (Equation 7.11) to emphasize its dependence on the cantilever's length:

$$\omega_{cant} = 3.52\sqrt{\frac{EI}{\rho AL^4}} \sim \frac{1}{L^2} \tag{7.24}$$

Thus, the natural or fundamental frequency is inversely proportional to the square of the beam's length—or of the building's height if it was modeled as a simple cantilever. But the literature features several empirically derived estimates of the fundamental frequency of a tall building as a function of L. For example, the fundamental frequency n_0 for braced steel frames and reinforced concrete shear wall buildings could be calculated as

$$\omega_0 = \frac{20\pi}{N} \tag{7.25}$$

where N is the number of stories in the building, which is essentially proportional to the building's height. Another study in which the frequencies of 163 buildings were measured produced this formula:

$$\omega_0 \cong \frac{289}{L} \quad (m) \tag{7.26}$$

Still another study noted a "widely used" formula that is particularly useful for reinforced concrete shear wall buildings and braced steel frames:

$$\omega_0 \cong \frac{69\sqrt{D}}{L} \quad (m) \tag{7.27}$$

where D is the building depth, with both L and D being measured in meters. Equation (7.27) is also said to be applicable to shear wall construction.

Finally, research has also suggested that the frequencies of steel moment-resisting frames and of reinforced concrete buildings would vary with length as, respectively,

$$\omega_0 \propto L^{-0.80} \quad \text{and} \quad \omega_0 \propto L^{-0.90} \tag{7.28}$$

Equations (7.25)–(7.28) all show the same (or similar) dependence: namely, frequency ~ $1/L$, which is in sharp contrast to the Euler–Bernoulli beam model predictions. Thus, it is natural to ask whether a tall building's frequency *can* be estimated from a beam theory calculation, or do the empirical formulas commonly used indicate a different behavior? Similarly, can a higher order beam theory, such as the well-known Timoshenko beam model, resolve this apparent conflict?

What is also interesting about these questions is that they are typically viewed in the context of extending the Euler–Bernoulli model of beam behavior, which does not admit shear deformation, to account for that shear response. But, as it turns out, we can incorporate shear behavior in two different ways and will, consequently, produce two different models. In one approach we invoke the Timoshenko beam model, in which we add the shear and bending deflections to produce the total beam deflection. As we observed in Chapter 6, we are thus adding the bending and shear stiffnesses together in *series*.

The second approach we could take would be to couple an Euler–Bernoulli beam to a shear beam in a *parallel* formulation in which each shares the same (total) deflection. We will see that this parallel coupling enables a model that allows us to see a frequency dependence on beam length (or building height) that is consistent with measured data. We will also see that the two models' behavior is heavily dependent on a dimensionless parameter relating a model's shear and bending stiffness (i.e., $\alpha^2 = SL^2/B$), although we will have to pay close attention to how we identify these stiffnesses in each of the two (i.e., Timoshenko and coupled Euler-to-shear) models.

Frequency-Height Dependence in Coupled Two-Beam Models

We now consider modeling a building as a vertically oriented cantilever beam in which the cross section is composed of two beams connected in parallel. For example, suppose we connected a symmetric external tube to a symmetric internal core, made of different materials, using axially rigid, mass-less transverse elements (Figure 7.1). The external tube is modeled as an elementary Euler–Bernoulli beam with a bending stiffness B and a center line transverse displacement of $w_b(x, t)$. The internal core has standard shear stiffness S, whose transverse displacement is $w_s(x, t)$. The rigid transverse connectors *ensure that the (beam) bending and (core) shear displacements are equal;* that is,

$$w_b(x,t) = w_s(x,t) \equiv w(x,t) \tag{7.29}$$

where $w(x, t)$ denotes the *common transverse displacement* of the coupled two-beam model of the tube and core.

We find the governing equation for this coupled beam system by applying Hamilton's principle to the appropriate total Lagrangian, L. The total Lagrangian is obtained as before by evaluating the difference between the

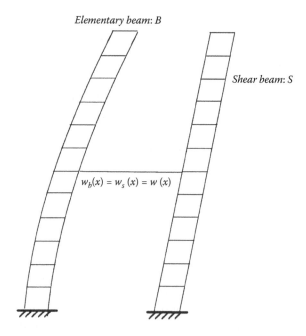

Elementary beam: B

Shear beam: S

$w_b(x) = w_s(x) = w(x)$

FIGURE 7.1
The continuous model of an Euler beam in bending B coupled in parallel with a pure shear S element. (See also Figure 6.3 for the discrete model.)

kinetic energy of the Euler and shear beams, $T_b + T_s$, and strain energy for the two beams, $U_b + U_s$:

$$\mathcal{L} = (T_b + T_s) - (U_b + U_s) \tag{7.30}$$

The kinetic energies for the elementary beam, T_b, and for the shear beam, T_s, are

$$T_b + T_s = \frac{\rho_b A_b}{2} \int_0^L \left(\frac{\partial w(x,t)}{\partial t} \right)^2 dx + \frac{\rho_s A_s}{2} \int_0^L \left(\frac{\partial w(x,t)}{\partial t} \right)^2 dx \tag{7.31}$$

The strain energies for the Euler–Bernoulli beam, U_b, and for the shear beam, U_s, are

$$U_b + U_s = \frac{B}{2} \int_0^L \left(\frac{\partial^2 w(x,t)}{\partial x^2} \right)^2 dx + \frac{S}{2} \int_0^L \left(\frac{\partial w(x,t)}{\partial x} \right)^2 dx \tag{7.32}$$

Then, combining Equations (7.30)–(7.32), we obtain the Lagrangian for the coupled two-beam model as, finally,

$$\mathcal{L} = \frac{\rho_b A_b}{2} \int_0^L \left(\frac{\partial w(x,t)}{\partial t} \right)^2 dx + \frac{\rho_s A_s}{2} \int_0^L \left(\frac{\partial w(x,t)}{\partial t} \right)^2 dx$$
$$- \left[\frac{B}{2} \int_0^L \left(\frac{\partial^2 w(x,t)}{\partial x^2} \right)^2 dx + \frac{S}{2} \int_0^L \left(\frac{\partial w(x,t)}{\partial x} \right)^2 dx \right] \tag{7.33}$$

We could again use the Lagrangian (Equation 7.34) in conjunction with Hamilton's principle to derive the equation of motion for the coupled composite beam, along with the appropriate boundary conditions. However, since our immediate interest is solely in finding a useful estimate of that beam's fundamental frequency, we use the Lagrangian (Equation 7.33) to develop a Rayleigh quotient for the beam by assuming the separable solution (Equation 7.7). When we substitute Equation (7.7) into the Lagrangian (Equation 7.33), we get the following equation for the fundamental frequency of the coupled Euler–shear model:

$$\omega_{ES}^2 = \frac{\dfrac{B}{2L^3} \int_0^1 (W''(\xi))^2 \, d\xi + \dfrac{S}{2L} \int_0^1 (W'(\xi))^2 \, d\xi}{\dfrac{\rho A L}{2} \int_0^1 (W(\xi))^2 \, d\xi} \tag{7.34}$$

Note that we have introduced into the quotient (Equation 7.34) a *total* specific mass ρA defined as

$$\rho A = \rho_b A_b + \rho_s A_s \tag{7.35}$$

Now we define two dimensionless coefficients, each of which is a simple number of order unity that depends solely on the mode shape $W(\xi)$ rather than on the mode shape amplitude or frequency:

$$\mu_{bending}^{ES} = \frac{\displaystyle\int_0^1 (W''(\xi))^2\, d\xi}{\displaystyle\int_0^1 (W(\xi))^2\, d\xi}\ ,\qquad \mu_{shear}^{ES} = \frac{\displaystyle\int_0^1 (W'(\xi))^2\, d\xi}{\displaystyle\int_0^1 (W(\xi))^2\, d\xi} \tag{7.36}$$

With the aid of the coefficients (Equation 7.36), we can write the Rayleigh quotient (Equation 7.34) as the sum of two distinct terms:

$$\omega_{ES}^2 = \left(\frac{B}{\rho A L^4}\right)\mu_{bending}^{ES} + \left(\frac{S}{\rho A L^2}\right)\mu_{shear}^{ES} \tag{7.37}$$

Equation (7.37) clearly shows that the fundamental frequency is proportional to the sum of two terms: The first incorporates bending and is inversely proportional to L^2 and the second term incorporates shear deformation and is inversely proportional to L. Then the true length dependence exhibited by the model clearly depends on the relative values of the ratios preceding each of the coefficients. If we now formally introduce the parameter α defined as

$$\alpha^2 = \frac{SL^2}{B} \tag{7.38}$$

then Equation (7.37) takes the form of

$$\omega_{ES}^2 = \left(\frac{B}{\rho A L^4}\right)\left[\mu_{bending}^{ES} + \alpha^2 \mu_{shear}^{ES}\right] \tag{7.39}$$

Clearly, then, for large values of α^2, the second term dominates and the fundamental frequency can be estimated to be

$$\omega_{ES}^2 \cong \left(\frac{B}{\rho A L^4}\right)\alpha^2 \mu_{shear}^{ES} = \frac{S}{\rho A L^2}\mu_{shear}^{ES} \tag{7.40}$$

Note that the frequency (Equation 7.40) decreases as $1/L$, which is consistent with the empirical results cited earlier. Thus, buildings that can be characterized as having large values of $\alpha^2 = SL^2/B$ can be modeled as coupled Euler–shear beams.

It is also interesting to note and easy enough to revise the preceding analysis to show that, had we wanted to account for P–Δ effects in accord with Equation (7.20), we would have defined an appropriate coefficient

$$\mu_{P-\Delta}^{ES} = \frac{\int_0^1 (1-\xi)(W'(\xi))^2\, d\xi}{\int_0^1 (W(\xi))^2\, d\xi} \tag{7.41}$$

and then found that the corresponding frequency would be given as

$$\omega_{ES}^2 = \left(\frac{B}{\rho AL^4}\right)\left[\mu_{bending}^{ES} + \alpha^2 \mu_{shear}^{ES}\right] - \left(\frac{g}{L}\right)\mu_{P-\Delta}^{ES} \tag{7.42}$$

So for the case where α^2 is large,

$$\omega_{ES}^2 \cong \left(\frac{S}{\rho AL^2}\right)\mu_{shear}^{ES} - \left(\frac{g}{L}\right)\mu_{P-\Delta}^{ES} \cong \left(\frac{S}{\rho AL^2}\right)\mu_{shear}^{ES} \tag{7.43}$$

That is, we are concluding that the P–Δ effects are, in fact, negligible. Why? Consider the ratio of the normal stress produced by the building's weight, $\rho g AL/A$, to the building's shear modulus, S/A. Since stresses are typically less than 1/1000 of a modulus, the P–Δ term can be neglected and Equation (7.43) is approximated accordingly.

It is important to keep in mind that the results presented here were derived based on the assumption that the *bending and shear deflections were one and the same*. This suggests an analogy to connecting the beam's (or building's) mass to two springs acting in *parallel*: one representing the bending response and the other the shear response.

Frequency-Height Dependence in Timoshenko Beams

We now briefly consider a single, vertically oriented cantilever beam with a uniform cross section that is symmetric through the thickness. (Again, we can easily extend this analysis to include nonuniform cross sections, without changing our basic result.) For this model, however, we use the Timoshenko beam model that incorporates both shear deformation and rotatory inertia, although the rotatory inertia effect is unimportant here because it is not a significant factor for tall, slender buildings. We denote the total transverse displacement of the beam's center line as $w(x, t)$, the bending rotation of line elements originally normal to the beam's center line as $\psi(x, t)$, and the rotation of the center line due to shear as $\beta(x, t)$. The formulation of the displacement field for this problem begins with the assumption that the *total slope of the beam is the sum of a component due to the bending rotation $\psi(x, t)$ and the shear rotation $\beta(x, t)$*:

$$\frac{\partial w(x,t)}{\partial x} = \psi(x,t) + \beta(x,t) \tag{7.44}$$

The corresponding displacement field is

$$u(x,y,z,t) = -z\psi(x,t) = -z\left(\frac{\partial w(x,t)}{\partial x} - \beta(x,t)\right)$$

$$v(x,y,z,t) = 0 \tag{7.45}$$

$$w(x,y,z,t) = w(x,t)$$

The corresponding engineering strains, axial and shear, are

$$\varepsilon_{xx}(x,z,t) = -z\frac{\partial \psi(x,t)}{\partial x} = -z\left(\frac{\partial^2 w(x,t)}{\partial x^2} - \frac{\partial \beta(x,t)}{\partial x}\right)$$

$$\gamma_{xz}(x,z,t) = \frac{\partial w(x,t)}{\partial x} - \psi(x,t) = \beta(x,t) \tag{7.46}$$

Note that the axial strain is linear through the beam's thickness and the shear strain is assumed to be constant. This is an unrealistic assumption whose effects will be compensated later through the introduction of a shear constant.

We can use Equation (7.46) to develop the Raleigh quotient for the Timoshenko beam in a process much like that which we presented before for the coupled two-beam model, and we would find a result that is remarkably similar in structure and appearance to that displayed in Equation (7.42). That is, including the P–Δ effect but ignoring rotatory inertia, the Timoshenko beam result is

$$\omega_{TB}^2 = \left(\frac{B}{\rho AL^4}\right)\left[\mu_{bending}^{TB} + \alpha^2\mu_{shear}^{TB}\right] - \left(\frac{g}{L}\right)\mu_{P-\Delta}^{TB} \tag{7.47}$$

However, the coefficients displayed in Equation (7.47) are, for the Timoshenko beam model, very different from their counterparts (Equation 7.36) for the Euler–shear model; that is,

$$\mu_{bending}^{TB} = \frac{\int_0^1 (\Psi'(\xi))^2 \, d\xi}{\int_0^1 (W(\xi))^2 \, d\xi}, \quad \mu_{shear}^{TB} = \frac{\int_0^1 (W'(\xi) - \Psi(\xi))^2 \, d\xi}{\int_0^1 (W(\xi))^2 \, d\xi}, \quad \mu_{P-\Delta}^{TB} = \frac{\int_0^1 (1-\xi)(W'(\xi))^2 \, d\xi}{\int_0^1 (W(\xi))^2 \, d\xi}$$

$$(7.48)$$

It is worth noting that the coefficients in Equation (7.48) become exactly those for the bending vibration of a self-loaded, elementary Euler–Bernoulli beam if we set $\Psi(\xi) = W'(\xi)$. Unfortunately, however, we cannot obtain the coupled two-beam model coefficients (Equation 7.36) from the Timoshenko results (Equation 7.48)—no doubt because the Timoshenko model is based on the assumption that the *total slope of the deflected beam is the sum of the bending and shear rotations,* which is analogous to a bending spring and a shear spring connected in *series* to the beam's (or building's) mass. (Remember that these elements are connected in parallel in the Euler–shear combination.)

So now we have two models—the Euler–shear beam combination and the Timoshenko beam—that appear to offer seemingly similar results: Note the similarity of Equations (7.42) and (7.47). But they also offer somewhat different results: Their respective coefficients (Equations 7.36 and 7.48) differ because their underlying displacement assumptions differ. Further, neither model conforms to the empirical results described earlier for all values of α^2. Is either model adequate or appropriate?

Now, in the case of Equations (7.36) and (7.41), the coefficients are independent of the mode amplitude because there is only a single degree of freedom in the Euler–shear model since the transverse displacement is clearly the same for both beams. On the other hand, the Timoshenko beam has two degrees of freedom, representing separately (if summed) the bending deformation and the shear deformation, so it produces a model that is only superficially similar: The Timoshenko coefficients (Equation 7.48) depend on the (modal) amplitudes, which removes the possibility of a straightforward analysis such as that we did for the coupled Euler–shear model.

In this context we also point out that, for the first mode of a cantilever shear beam, for which $W(\xi) \sim \sin(\pi\xi/2)$, we can use Equation (7.36) to calculate $\mu_{shear}^{ES} = \pi^2/4$ and Equation (7.40) to calculate a classical shear beam result:

$$\omega_{ES}^2(\alpha^2 \gg 1) \Rightarrow \left(\frac{\pi}{2}\right)^2 \left(\frac{G}{\rho L^2}\right)$$

$$(7.49)$$

Here it can be seen that Equations (7.43) and (7.49) both conform to the empirical data outlined in the first section; that is, $\omega \sim 1/L$. Thus, it seems clear that the first mode of a tall building can be estimated as if it were a shear beam.

We further confirm this conclusion by noting that the result (Equation 7.49) can be used to calculate the speed of a shear wave in the beam (building) as a function of frequency. In particular, Equation (7.49) can be solved for the shear wave speed c_s expressed as the square root of the modulus-to-density ratio so that

$$c_s = \sqrt{\frac{G}{\rho}} = \left(\frac{2}{\pi}\right)\omega_{ES}L \tag{7.50}$$

Now, using the empirical results reviewed earlier in the chapter, we can explicitly calculate the shear wave speeds. For example, by using the estimate Equation (7.26) we get a shear speed of 184 m/s. Further, by using estimate Equation (7.25) with the additional assumption that each story is 3.5 m high, we find a shear speed of 140 m/s. These values are consistent with previously reported estimates of the shear wave speed.

Comparing Frequencies for Coupled Two-Beam Models

We now find two solutions for the fundamental frequency of the Euler–shear coupled system: an exact solution obtained in the traditional manner and a set of approximate estimates obtained by using different mode shapes in the system's Rayleigh quotient. For the exact solution, we apply Hamilton's principle to the Lagrangian (Equation 7.32) to get the equation of motion,

$$B\frac{\partial^4 w(x,t)}{\partial x^4} - S\frac{\partial^2 w(x,t)}{\partial x^2} + (\rho_b A_b + \rho_s A_s)\frac{\partial^2 w(x,t)}{\partial t^2} = 0 \tag{7.51}$$

as well as the corresponding boundary conditions, which, for a cantilever fixed at $x = 0$ and free of moment and shear at the tip $(x = L)$, are

$$w(0,t) = 0, \quad \frac{\partial w(0,t)}{\partial x} = 0, \quad M(L,t) = -B\frac{\partial^2 w(L,t)}{\partial x^2} = 0$$

$$V(L) = -B\frac{\partial^3 w(L,t)}{\partial x^3} + S\frac{\partial w(L,t)}{\partial x} = 0 \tag{7.52}$$

For harmonic motion, per Equation (7.7), the equation of motion becomes an eigenvalue problem for the resonant frequencies ω:

$$W^{iv}(x) - (S/B)W''(x) - (\rho A\omega^2/B)W(x) = 0 \tag{7.53}$$

Since Equation (7.53) has constant coefficients, we can write its solution as

$$W(x) = e^{px/L} \tag{7.54}$$

which leads us to the corresponding characteristic equation:

$$p^4 - \alpha^2 p^2 - (\omega/\omega_0)^2 = 0 \tag{7.55}$$

Recall that we defined α in Equation (7.38) and note that the factor ω_0 is given by

$$\omega_0 = \sqrt{B/\rho AL^4} \tag{7.56}$$

Now, Equation (7.55) is a biquadratic equation that we can easily solve to find

$$p^2 = \frac{\alpha^2}{2}\left(1 \pm \sqrt{1 + \left(\frac{2\omega}{\alpha^2\omega_0}\right)^2}\right) \tag{7.57}$$

If we inspect Equation (7.57) closely, we see that the nature of the roots changes markedly around the value "1" because of the plus-or-minus option preceding the radicand. We can capitalize on that recognition to clarify our solution by completing the square of that radicand. So, instead of seeking values of ω, we will search instead for values of a new variable, η, which we define as

$$\left(\frac{\omega}{\alpha^2\omega_0}\right)^2 = \eta^2(1 + \eta^2) \tag{7.58}$$

Then, the characteristic Equation (7.57) becomes

$$p^2 = \frac{\alpha^2}{2}(1 \pm (1 + 2\eta^2)) \tag{7.59}$$

Equation (7.59) has four roots:

$$p_{1,2}^2 = -\alpha^2 \eta^2, \quad p_{3,4}^2 = \alpha^2(1+\eta^2) \tag{7.60}$$

We see in Equation (7.60) the significant change in the roots that we anticipated—namely, that two will be purely imaginary and two purely real. This means that we can write the homogeneous solution to Equation (7.53) as

$$W(x) = C_1 \sin \alpha \eta \xi + C_2 \cos \alpha \eta \xi + C_3 \sinh \alpha \sqrt{1+\eta^2} \xi + C_4 \cosh \alpha \sqrt{1+\eta^2} \xi \tag{7.61}$$

We now use the boundary conditions (Equation 7.52) to establish a sequence of homogeneous equations for the unknown constants C_i. From the deflection and slope conditions at the cantilever base ($x = \xi = 0$), we find that

$$C_2 + C_4 = 0 \quad \text{and} \quad \alpha \eta C_1 + \alpha \sqrt{1+\eta^2} C_3 = 0 \tag{7.62}$$

Equation (7.62) allows us to recast the remaining two boundary conditions entirely in terms of the constants C_1 and C_2. From the moment condition at the tip (Equation 7.52), we get

$$\frac{C_2}{C_1} = -\frac{\eta^2 \sin \alpha \eta + \eta \sqrt{1+\eta^2} \sinh \alpha \sqrt{1+\eta^2}}{\eta^2 \cos \alpha \eta + (1+\eta^2) \cosh \alpha \sqrt{1+\eta^2}} \tag{7.63}$$

From the shear condition at the tip, we find the following:

$$\frac{C_2}{C_1} = \frac{(1+\eta^2) \cos \alpha \eta + \eta^2 \cosh \alpha \sqrt{1+\eta^2}}{(1+\eta^2) \sin \alpha \eta - \eta \sqrt{1+\eta^2} \sinh \alpha \sqrt{1+\eta^2}} \tag{7.64}$$

When we set Equations (7.63) and (7.64) equal one to another, we get (after some more algebra!) our expected transcendental equation for the frequency variable η:

$$2 + \frac{(1+2\eta^2 + 2\eta^4) \cos \alpha \eta \cosh \alpha \sqrt{1+\eta^2}}{\eta^2(1+\eta^2)} + \frac{\sin \alpha \eta \sinh \alpha \sqrt{1+\eta^2}}{\eta \sqrt{1+\eta^2}} = 0 \tag{7.65}$$

In principle, we have in Equation (7.65) the exact solution for the free vibration frequencies of a cantilevered coupled Euler–shear system, which is clearly not terribly useful for back of the envelope estimates! It is also worth noting that we have potential problems even obtaining a numerical solution to Equation (7.65): We have no idea where to start; that is, how do we establish starting values of η? And how do we deal with the apparent singularity created at $\eta = 0$, especially when we might want to consider zero values for the shear-bending parameter α? It turns out that we can easily resolve the numerical issues by introducing a new variable,

$$\gamma = \alpha\eta \tag{7.66}$$

With the aid of the transformation (Equation 7.66), we can rewrite our transcendental frequency equation as

$$2 + \frac{(\alpha^4 + 2\alpha^2\gamma^2 + 2\gamma^4)\cos\gamma\cosh\sqrt{\alpha^2 + \gamma^2}}{\gamma^2(\alpha^2 + \gamma^2)} + \frac{\alpha^2\sin\gamma\sinh\sqrt{\alpha^2 + \gamma^2}}{\gamma\sqrt{\alpha^2 + \gamma^2}} = 0 \tag{7.67}$$

In Table 7.1 we show (and will further discuss next) some results for the frequency as a function of α.

The exact solution for the fundamental frequency is clearly sufficiently complicated that it has little value for quick, "back of the envelope" calculations. To see how well a Rayleigh quotient would serve, we see how the mode shape used earlier for elementary beams would work for the Euler–shear coupled system. So, for the mode shape

$$W_1(x) = W_{01}\left(1 - \cos\frac{\pi x}{2L}\right) \tag{7.68}$$

TABLE 7.1

Exact Values and Estimates of the Dimensionless Frequency of a Cantilevered Coupled Euler–Shear System for a Range of Values of the Parameter α

α	η	$\gamma = \alpha\eta$	$\sqrt{\dfrac{\rho AL^4}{B}}\omega_{exact}^{ES}$	$\sqrt{\dfrac{\rho AL^4}{B}}\omega_1^{ES}$	$\sqrt{\dfrac{\rho AL^4}{B}}\omega_2^{ES}$	$\sqrt{\dfrac{\rho AL^4}{B}}\omega_3^{ES}$
0	∞	1.875	3.52	3.66	3.53	4.47
1.0	1.908	1.908	4.11	4.35	4.11	5.16
2.0	0.972	1.944	5.42	5.94	5.50	6.83
3.0	0.646	1.938	6.92	7.92	7.25	8.95
4.0	0.476	1.906	8.44	10.04	9.15	11.26
5.0	0.373	1.867	9.96	12.25	11.12	13.67

we would find that the Rayleigh quotient (Equation 7.39) yields the result

$$\omega_1^{ES} = \left(\frac{(3.67)^2 B}{\rho A L^4} \right)^{1/2} \sqrt{(1+0.408\alpha^2)} = 1.04\omega_{cant} \sqrt{1+0.408\alpha^2} \qquad (7.69)$$

For the mode shape

$$W_2(x) = W_{02} \left[\left(\frac{x}{L} \right)^4 - 4\left(\frac{x}{L} \right)^3 + 6\left(\frac{x}{L} \right)^2 \right] \qquad (7.70)$$

we find

$$\omega_2^{ES} = \left(\frac{(3.53)^2 B}{\rho A L^4} \right)^{1/2} \sqrt{(1+0.357\alpha^2)} = 1.00\omega_{cant} \sqrt{1+0.357\alpha^2} \qquad (7.71)$$

and for the mode shape

$$W_3(x) = W_{03} \left(\frac{x}{L} \right)^2 \qquad (7.72)$$

we find

$$\omega_3^{ES} = \left(\frac{(4.47)^2 B}{\rho A L^4} \right)^{1/2} \sqrt{(1+0.334\alpha^2)} = 1.27\omega_{cant} \sqrt{1+0.334\alpha^2} \qquad (7.73)$$

In Table 7.1 we show the values obtained for the fundamental frequencies from the exact result (Equation 7.67) and for the three Rayleigh quotients (Equations 7.69, 7.71, and 7.73).

Conclusions

We have used this chapter to illustrate the power of a classic mechanics tool, the Rayleigh quotient. First, we showed how it could be used very easily to get very accurate estimates of the fundamental frequency of a simple cantilever. We went on to show how we could also use a Rayleigh quotient, together with some thinking about dimensions, to investigate how different models—a Timoshenko beam and a coupled Euler–shear beam system—match up with empirical measurements of the fundamental frequencies of tall buildings. It turned out that while both models gave the same results superficially (see Equations 7.42 and 7.47), a coupled Euler–shear model is easier to use because it depends only on a single mode shape, while the Timoshenko beam requires two: the (independent) bending rotation and the shear deflection.

Further, the kinematic behavior assumed in that coupled Euler–shear model also seems to fit the physics (e.g., compare tube-and-core construction with the parallel nature of the two-beam model in which transverse displacements due to bending and to shear are identical). We concluded the chapter by calculating both exact and approximate results for the fundamental frequency of a cantilevered Euler–shear system, once again showing how a Rayleigh quotient produces quick, accurate estimates of that frequency over a practical range of values of the building parameter $\alpha^2 = SL^2/B$.

Bibliography

Balendra, T. 1984. Free vibration of a shear wall-frame building on an elastic foundation. *Journal of Sound and Vibration* 96 (4): 437–446.

Basu, A. K., A. K. Nagpal, and S. Kaul. 1984. Charts for seismic design of frame-wall systems. *Journal of Structural Engineering* 110 (1): 31–46.

Basu, A. K., A. K. Nagpal, and A. K. Nagar. 1982. Dynamic characteristics of frame-wall systems. *Journal of the Structural Division, ASCE* 108 (ST6): 1201–1218.

Dym, C. L. 2002. *Stability theory and its applications to structural mechanics.* Mineola, NY: Dover Publications.

Dym, C. L., and I. H. Shames. 1973. *Variational methods in solid mechanics.* New York: McGraw-Hill.

Dym, C. L., and H. E. Williams. 2007. Estimating fundamental frequencies of tall buildings. *Journal of Structural Engineering* 133 (10): 1479–1483.

Ellis, B. R. 1980. An assessment of the accuracy of predicting the fundamental natural frequencies of buildings and the implications concerning the dynamic analysis of structures. *Proceedings of the Institution of Structural Engineers* 69 (2): 763–776.

Goel, R. K., and A. K. Chopra. 1997. Period formulas for moment resisting frame structures. *Journal of Structural Engineering* 123 (11): 1454–1461.

———. 1998. Period formulas for concrete shear wall buildings. *Journal of Structural Engineering* 124 (4): 426–433.

Heidebrecht, A. C., and B. Stafford Smith. 1973. Approximate analysis of tall wall-frame structures. *Journal of the Structural Division, ASCE* 99 (ST2): 199–221.

Lee, L.-H., K.-K. Chang, and Y.-S. Chun. 2000. Experimental formula for the fundamental period of RC buildings with shear-wall dominant systems. *Structural Design of Tall Buildings* 9: 295–307.

Miranda, E. 1999. Approximate seismic lateral deformation demands in multistory buildings. *Journal of Structural Engineering* 125 (4): 417–425.

Miranda, E., and C. J. Reyes. 2002. Approximate lateral drift demands in multistory buildings with nonuniform stiffness. *Journal of Structural Engineering* 128 (7): 840–849.

Miranda, E., and S. Taghavi. 2005. Approximate floor acceleration demands in multistory buildings. I: Formulation. *Journal of Structural Engineering* 131 (2): 203–211.

Newmark, N. M., and W. J. Hall. 1981. *Earthquake spectra and design*. Berkeley, CA: Earthquake Engineering Research Institute.

Rahgozar, R., H. Safari, and P. Kaviani. 2004. Free vibration of tall buildings using Timoshenko beams with variable cross-section. In *Structures under shock and impact VIII*, ed. N. Jones and C. A. Brebbia. Ashurst, New Forest, England: WIT Press.

Safak, E. 1999. Wave propagation formulation of seismic response of multistory buildings. *Journal of Structural Engineering* 125 (4): 426–437.

Stafford Smith, B., and A. Coull. 1991. *Tall building structures: Analysis and design*. New York: John Wiley & Sons.

Taranath, B. S. 1988. *Structural analysis and design of tall buildings*. New York: McGraw-Hill.

Williams, H. E. 2008. An asymptotic solution of the governing equation for the natural frequencies of the cantilevered, coupled beam model. *Journal of Sound and Vibration* 312: 354–359.

Problems

7.1   Use a Rayleigh quotient to estimate the natural frequency ω of a simply supported Euler–Bernoulli beam. Explain why that estimate is (or is not) a good estimate.

7.2   Use a Rayleigh quotient to estimate the natural frequency ω of a simply supported Euler–Bernoulli beam with a concentrated mass M at a distance x_0 from the left support ($0 \leq x_0 \leq L$). Explain why that estimate is (or is not) a good estimate.

7.3   Use a Rayleigh quotient to estimate the natural frequency ω of a cantilevered Euler–Bernoulli beam with a concentrated mass M at a distance x_0 from the base ($0 \leq x_0 \leq L$). Explain why that estimate is (or is not) a good estimate.

7.4   Use a Rayleigh quotient to determine the buckling load P of a simply supported Euler column. Explain why that estimate is (or is not) a good estimate.

7.5   Use a Rayleigh quotient to determine the buckling load P of a cantilevered Euler column. Explain why that estimate is (or is not) a good estimate.

7.6   Use a Rayleigh quotient to determine the buckling load P of an Euler column that is simply supported at $x = 0$ and clamped at $x = L$. Explain why that estimate is (or is not) a good estimate.

7.7   Equation (7.66) makes it clear that $\gamma = \gamma(\alpha)$. Using this insight, show that the Euler–shear system transcendental Equation (Equation

7.67) reduces to the corresponding Euler–Bernoulli beam result (Equation 7.10) as $\alpha \to 0$.

7.8 Using Equation (7.66), show that Equation (7.58) can be written as

$$(\omega(\gamma, \alpha))^2 = \gamma^2 \left[1 + \left(\frac{\gamma}{\alpha} \right)^2 \right] \frac{S}{\rho A L^2}$$

7.9 Assuming that γ ($\alpha \to \infty$) is finite, show that the result obtained in Problem 7.8 produces the limit

$$(\omega(\alpha \to \infty))^2 \to \gamma^2 \left(\frac{S}{\rho A L^2} \right)$$

What kind of beam behavior does this result suggest?

7.10 Define a new variable $\varepsilon = 1/\alpha$ for the case of $\alpha \gg 1$ and expand $\gamma(\varepsilon)$ in the power series (or perturbation expansion)

$$\gamma(\varepsilon) = \Omega_0 (1 + a_1 \varepsilon + a_2 \varepsilon^2 + \cdots)$$

where the $a_i \neq a_i(\varepsilon)$; that is, the a_i are independent of ε. Substitute this expansion into Equation (7.67) and show that $\Omega_0 = (2n - 1)\pi/2$ for $n = 1, 2, \ldots$ and $a_1 = a_2 = 1$ so that

$$\gamma^2 (1 + \varepsilon^2 \gamma^2) = \Omega_0^2 (1 + 2\varepsilon + (3 + \Omega_0^2)\varepsilon^2 + \cdots)$$

Index

Milton Keynes UK
Ingram Content Group UK Ltd.
UKHW040059071024
449327UK00019B/677